FOSSILS OF THE MILWAUKEE FORMATION:

A DIVERSE MIDDLE DEVONIAN BIOTA FROM WISCONSIN, USA

Kenneth C. Gass, Joanne Kluessendorf, Donald G. Mikulic
and Carlton E. Brett

SSP
SIRI SCIENTIFIC PRESS

Authors, if you would like to publish with us please email your proposed titles (at any stage of preparation) to:
books@siriscientificpress.co.uk
or contact us via our website

(we specialize in high quality, rapid production and short print run titles that would tend to be overlooked by larger publishers)

ISBN 978-0-9957496-7-2

Published by Siri Scientific Press, Manchester, UK

This and related titles are available directly from the publisher at

http://www.siriscientificpress.co.uk

Printed and bound in the UK

In Memory of
Dr. Joanne Kluessendorf
(1949–2018)

Joanne Kluessendorf with cast of *Dunkleosteus*; December, 2017.
Courtesy of Beth A. Johnson and The Geological Society of America.

1: Lycopod (plant)
2: *Prototaxites milwaukeensis* (fungus)
3: Cladoxylopsid? log
4: *Desmograptus* sp. (dendroid graptolite)
5: *Conularia milwaukeensis* (conulariid)
6: *Onychodus* sp. (sarcopterygian)
7: *Eastmanosteus pustulosus* (arthrodire)
8: *Cladodus monroei* (shark)
9: *Ptyctodus ferox* (ptyctodont)
10. Ammonoid (cephalopod)
11: *Acleistoceras? wisconsinense* (nautiloid)
12: *Acleistoceras? calvini* (nautiloid)
13: *Gyronaedyceras eryx* (nautiloid)
14: *Melocrinites subglobosus* (crinoid)
15: *Melocrinites nodosus spinosus* (crinoid)
16: *Taxocrinus telleri* (crinoid)
17: *Melocrinites?* sp.
18: *Nucleocrinus obovatus* (blastoid)
19: *Leioclema minutissimum* and other bryozoans
20. *Fistulipora* sp. (bryozoan)
21: *Eridotrypa appressa* (bryozoan)
22: *Sulcoretepora hamiltonensis* (bryozoan)
23: *Odontophyllum* cf. *O. convergens* (rugose coral)
24: *Aulopora* cf. *A. annectens* (tabulate coral)
25: *Devonatrypa?* sp. (brachiopod)
26: *Strophodonta halli* (brachiopod)
27: *Cranaena* sp. (brachiopod)
28: *Pseudodechenella?* sp. (trilobite)
29: *Eldredgeops? norwoodensis* (trilobite)
30: *Palaeoneilo* spp. (bivalves)
31: *Krama?* sp. (edrioasteroid)
32: *Microconchus* sp. (tentaculitoid)

Seascape/landscape of Middle Devonian Milwaukee showing some of its typical inhabitants and their probable modes of life, based on fossils from the Milwaukee Formation. Not shown are about 220 other known named species, un-named species, subspecies, and ichnospecies. ©Richard Bizley

Cover images (left to right, top to bottom): *Tentaculites bellulus* (tentaculitoid); seascape/landscape of Middle Devonian Milwaukee, Wisconsin (acrylic painting, ©Richard Bizley); *Taxocrinus telleri* (crinoid); *Gyronaedyceras eryx* (nautiloid); *Eldredgeops norwoodensis* (trilobite); *Melocrinites nodosus spinosus* (crinoid); lycopod branch (plant); *Protoleptostrophia serrata* (brachiopod); *Eridotrypa appressa* (bryozoan).

BIZLEY

Preface

Over one hundred years have passed since H.F. Cleland wrote *The Fossils and Stratigraphy of the Middle Devonic of Wisconsin* (1911) — the first and only volume written largely on the Milwaukee Formation biota. Typical of similar works of its day, its illustrations were mostly black and white drawings. During the last century, new material has been gathered, many additional species have come to light, much has been learned, and advances in technology have made it possible to economically publish books having many full color photographs. Today around 250 species, subspecies, unnamed species and ichnospecies have been identified from the Milwaukee Formation, nearly 180 of which are illustrated in this book. These are represented by new and old material, including type and figured specimens wherever practical. Moreover, the large collections from this diverse biota have enabled a wide variety of fossil studies that depend on large sample sizes. It is our hope that this compilation of information and photographs will enlighten and entertain regional fossil collectors, spark the interest of researchers and budding paleontologists, and reach the hands of graduate students and professionals studying Devonian paleontology and others seeking fertile grounds on which to base their research, as offered in this much-neglected but diverse Middle Devonian biota.

Each major taxon of the biota is covered in its own chapter. Each of those chapters includes a discussion of that taxon and in most cases photographs of representative Milwaukee Formation fossils labeled to show important observable morphological features. The chapter then presents photographs of many of the species identified within that major taxon. The names of the other species of that taxon known to exist in this formation but not shown are listed. Terms that are introduced but not defined in the text are defined in the Glossary. Sections on stratigraphy, Milwaukee's cement industry, paleoenvironments, modes of life, organism interactions, taphonomy, fossil collecting in the Milwaukee Formation, preparation methods, and major collections round out the discussion.

Whenever known, the collectors of the fossils shown are indicated with their initials in the figure captions. The identities of those collectors are as follows: Benedict Bagrowski (BB), Elmer Dahlman (ED), Ira Edwards (IE), Marlene O. Emielity (MOE), Damien K. Gass (DKG), Jeanette L. Gass (JLG), Joshua S. Gass (JSG), Kenneth C. Gass (KCG), Kevin M. Gass (KMG), Lauri A. Gass (LAG), Todd C. Gass (TCG), Thomas A. Greene (TAG), Spence Havlick (SH), Thomas James (TJ), Ronald Kneusel (RK), Thomas Maier (TM), Donald G. Mikulic (DGM), Milwaukee Cement Company (MCC), Charles E. Monroe (CEM), Gilbert O. Raasch (GOR), Jeannette E. Roberts (JER), H. Rollins (HR), Lindsey V. Stenerson (LVS), Karly S. Tellekson (KST), Edgar E. Teller (EET), and Stanley Wisniewski (SW). The collectors of the specimens residing in the Greene Geological Museum were typically not recorded in the museum catalogs, but were either Thomas A. Greene, various Milwaukee Cement Company workers, or others.

Classification, descriptions and morphological terminology for invertebrates generally follow the current editions of the appropriate volumes and parts of the *Treatise on Invertebrate Paleontology*. *Fishes of the World, Fifth Edition* (Nelson et al., 2016) was similarly an indispensable resource for fishes. The generic placement of many species has been updated based on the most recently published studies and consultation with various specialists. Identifications should be considered tentative and targets for future investigation. Due to the tentative nature of the identifications, diagnoses are not given.

Prefixes of specimen numbers given in the figure captions indicate the respective repositories, as follows: MPM=Milwaukee Public Museum; UWM-GM=Greene Geological Museum, University of Wisconsin—Milwaukee; V or PAL=National Museum of Natural History, Washington, D.C.; VPF=Museum of Comparative Zoology, Harvard University; WESM=Weis Earth Science Museum, Menasha, Wisconsin. The prefix XXX-KCG indicates that the specimen resides at the indicated museum, but the catalog number is that given by KCG and not by the museum. Higher taxonomic placement of the species shown in each photograph, typically order and family, are stated in the corresponding figure caption (e.g., Arthrodiriformes:

Dunkleosteidae). If the figure caption does not state where the fossil was found, the specimen was collected either at one of the quarries near the Milwaukee River in northeastern Milwaukee County (mainly those operated by the Milwaukee Cement Company), or from Estabrook or Lincoln Park. Unless otherwise indicated, the photographic work was that of KCG.

KCG, DGM and JK are indebted to the late Joseph G. Emielity, Katherine G. Nelson, and Gilbert O. Raasch—each being curators of Milwaukee Formation fossils—for their friendship, for sharing their wealth of knowledge, and for solidifying our interest in this subject. Richard Fortey is thanked for reviewing a draft of the manuscript, especially the introductory and arthropod chapters, and providing useful input. We thank the following for reading drafts of the section on the indicated group of organisms and/or examining photographs of that group and providing valuable input: Robert K. Carr (fishes); Brian Chatterton and James Cook (trilobites); Jed Day (brachiopods); Andrej Ernst (bryozoans); Paul Kenrick (land plants); David Loydell (graptolites); George McIntosh (blastoids); Colin Sumrall (edrioasteroids); Heyo Van Iten (conulariids); Olev Vinn (tentaculitoids); and Brian Witzke (crinoids). A special thank you goes to Patricia Coorough-Burke for open access to the material at the Milwaukee Public Museum and providing much information and support. We are indebted to Stephen Dornbos and his staff (Greene Geological Museum) and the late Joanne Kluessendorf (Weis Earth Science Museum) for access to the collections in their respective museums and accommodating our needs. Amanda Millhouse and Kathy Hollis (National Museum of Natural History) and Jessica Cundiff (Museum of Comparative Zoology) are thanked for photographing material in their respective museums. Linda Deith (The Wisconsin Geological and Natural History Survey), Emma MacPherson (Canadian Society of Petroleum Geologists), and Beth A. Johnson are thanked for permission to use the indicated published images. Robert K. Carr, Terry C. Grande, and Joanne Kluessendorf are thanked for allowing us to publish figures based on their drawings. Lisa Marine (Wisconsin Historical Society), Jennifer Pahl (Milwaukee Public Library), and Susan Peschel (American Geographical Society Library, University of Wisconsin-Milwaukee Libraries) are thanked for providing scans of images and other materials in their care. JK and DGM thank the Wisconsin Historical Society for use of their archives. We also thank Denise Baker for help with Word issues; Richard Bizley for painting the cover image; David Bradbury for helping locate specimens; David Comfort for preparing certain fossils; Niles Eldredge, Gerald Kloc, and Bruce Lieberman for helpful discussions on phacopid systematics; Curt Klug for years of collecting trips with DGM and JK; Todd Gass for providing hardware and software support and for helping process some of the images; Ron Kneusel for suggesting that the book be written; J. Maisey for verifying the identity of the shark fin spine; and Dietmer Schumacher for providing conodont preparation methods and permission to use his published images. Finally, we thank David Penney of Siri Scientific Press for extending the opportunity to publish this work and for making improvements.

Contents

Introduction

Wisconsin's Devonian rocks are so limited in extent that they are often glossed over or bypassed entirely by both fossil collectors and students of regional paleontology and stratigraphy. Some geologic maps do not even indicate the presence of Devonian formations in the state. This holds true of even Wisconsin's most diverse and fossiliferous sequence of Devonian strata—the Milwaukee Formation.

Yet in spite of its recent relative obscurity, as compared to the Devonian strata of other states, the Milwaukee Formation has produced an abundance of fossils from a diverse mix of invertebrates, fishes, microorganisms, land plants, and other organisms, as well as several forms of trace fossils, with some groups having been quite well known to researchers of the Devonian biota in the late 19th and early 20th centuries. The bulk of these fossils come from a series of quarries in the vicinity of Milwaukee (see Chapter 1) and two parks that now essentially occupy their space. A few nearby exposures, well cores and material from tunnel excavations provided additional informative fossils (Fig.1).

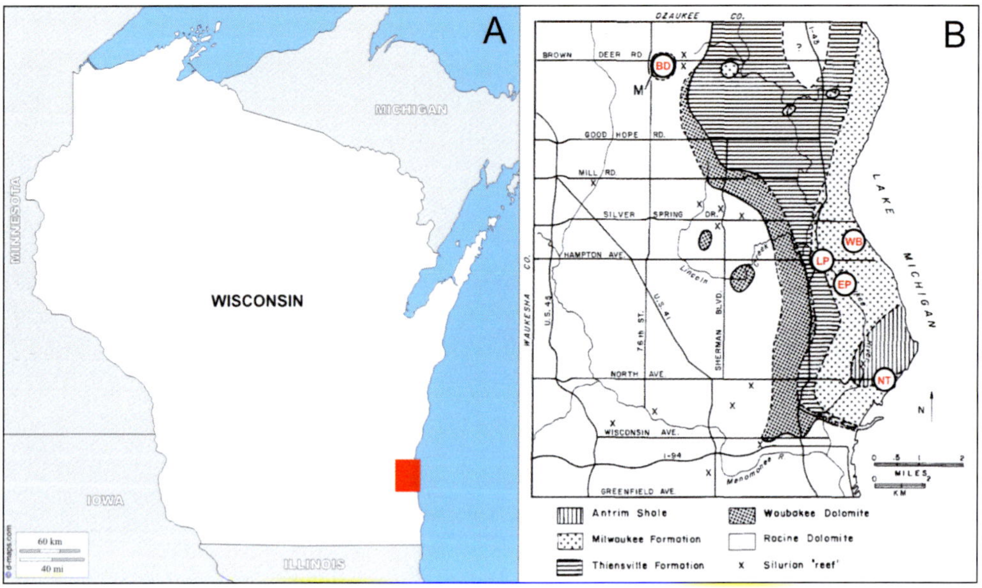

Fig. 1. Maps showing the area under study. **A:** Large scale map of Wisconsin and surrounding states. The red rectangle marks the location of the region shown in B. **B:** Bedrock geology map of northern half of Milwaukee County, Wisconsin. The locations of important collecting sites are also indicated. BD=Brown Deer; EP=Estabrook Park; LP=Lincoln Park; NT=North Avenue Intake Tunnel; WB and M="Whitefish Bay" of Chamberlin, a small outlier of Milwaukee Formation described by Chamberlin (1877) and Raasch (1925, unpublished field notes). Estabrook and Lincoln Parks also mark the approximate sites of the quarries of the Milwaukee Cement Company and several smaller cement companies. From Mikulic & Kluessendorf, 1988a, Fig. 1. Courtesy of Wisconsin Geological and Natural History Survey. Locality markers added by Todd Gass.

Along with the sources mentioned above, isolated fossils from the Milwaukee Formation scraped loose by the Pleistocene glaciers that scoured the region are spread sparsely around the southeastern part of Wisconsin and mixed with rocks from older formations. It was those brachiopods, crinoid pieces and other invertebrate fossils that caught the attention of Gass and Mikulic in the late 1950s. In the Milwaukee area, Silurian fossils far outnumber those from other periods; yet some of the Devonian fossils have a special appeal. Most of Wisconsin's Silurian fossils and many of the Devonian ones are preserved only as molds of the interior and exterior surfaces of shells, thus lacking surface details making identification difficult, and giving only an approximation of the organisms' true appearance. However, many of the fossils from the Milwaukee Formation are composed of the mineralized hard parts themselves, preserving external surface details. This good state of preservation of shelled and some non-shelled organisms plus contrasting colors, abundance, and great variety of these Devonian fossils make it easier to identify the organisms and imagine what life must have looked like around 390 million years ago. Furthermore, its broad array of invertebrates, together with the preservation of terrestrial plants, giant fungi, and many species of fish, make the Milwaukee Formation biota one of North America's richest and most diverse biotas coming from a single Devonian formation. The fossils representing this diversity are the subject of this book.

Most research on the taxonomy of the Milwaukee Formation biota was undertaken during the late 1800's and early 1900's. Beginning with Lapham's first study of these fossils in the 1840's there have been many descriptions of various aspects of this biota including systematic studies of specific fossil groups, biostratigraphic studies, and paleoenvironmental research. The first significant overview of the entire biota was presented by Whitfield (1882) as part of the 1873–1879 Geological Survey of Wisconsin primarily under the direction of T.C. Chamberlin. Whitfield stated that the purpose of his report was not to provide a complete representation of the fauna but only to describe "a few of the more prominent known species, together with some of the most characteristic new ones." He did record all of the taxa that were known at the time from the "Hamilton" beds in his *General List of the Fossils*, (p. 350–363), where 37 taxa were reported from the Hamilton beds. Cleland (1911) greatly expanded the number of taxa known from these beds to 160 taxa. Since then Milwaukee Formation taxa have been the subject of many local and regional studies of Midwestern Devonian fossils. This book is the first comprehensive overview of its entire biota in more than one hundred years.

Chapter 1 — Cementing its Place in History

In the mid-1840s Increase A. Lapham was the first person on record to have collected fossils from what would later be known as the Milwaukee Formation (Mikulic, 1983). A prominent citizen of early Milwaukee and a true Renaissance man, Lapham was also that city's first scientist. He wrote Wisconsin's first scientific paper and wrote the first book published in Wisconsin. It was largely at his urgings that the National Weather Service would eventually be established (Bergland & Hayes, 2014). It was also Lapham who alerted the eminent geologist James Hall to the fish dermal plates and spines that he found in the Milwaukee Formation's strata, which led to the strata's assignment to the Devonian.

Lapham recognized the similarity of this Devonian rock to that used in the natural cement industry (Kluessendorf & Mikulic, 2008), but it wasn't until 1873 that commercial exploitation of the rock was seriously considered. In that year Joseph R. Berthelet conducted tests on the rock and determined that it was indeed well suited for making natural cement (Greene & Berthelet, 1949). In 1875 he established the Milwaukee Cement Company (MCC) and production began the following year (Fig. 2). MCC's natural cement manufacturing operations were considered state of the art for the industry (Lewis, 1900). The product was regarded by some engineers as the best hydraulic rock cement in the world (Anonymous, 1893) due to its superior physical properties (Griffin, 1890). By 1891 the MCC was the largest producer of natural cement in the country, with the product being shipped to many states and used to build numerous important structures, such as bridges across the Mississippi and Missouri Rivers (Shorewood Historical Society, 2016).

During this time conditions could not have been better for discovering fossils in the Milwaukee Formation. This is mostly due to the huge volume of rock that was chopped, trimmed and lifted by hand, and to the "gentlemen paleontologists" who would pay the quarry workers generously for their best fossil finds (Mikulic, 1983). These self-educated naturalists included the pharmacist Thomas A. Greene, businessman Edgar E. Teller, and attorney Charles E. Monroe. Greene eventually became a board member and Vice President of the MCC (Greene & Berthelet, 1949). His collection, including fossils from the Milwaukee Formation, together with the museum that housed it, was granted National Historic Landmark status in 1993 (National Park Service, 2016). The majority of the fossils shown in this book were from the collections of those three men.

By 1909 natural cement could no longer compete in price with Portland cement, and the productive layers at the MCC quarries had been mostly quarried away (Kluessendorf & Mikulic, 2008). The decision was thus made to end quarrying operations. Filling of the quarries and construction of Estabrook Park at the spot previously occupied by some of them began by 1916. The remaining quarries on both sides of the Milwaukee River were eventually filled and replaced by Lincoln Park, radio and television transmission towers, parking lots, and other structures. Consequently, the thousands of fossil specimens that Greene, Teller and Monroe acquired from these quarries will probably remain unmatched in quality and abundance due to the nature of modern quarrying methods and scarcity of exposed strata. Fortunately, many of the large collections made while the Milwaukee Cement Company quarries were still active, those made by Gilbert O. Raasch before the quarries were filled, and those assembled at Estabrook and Lincoln Parks by Joseph G. Emielity and many others remain available for study in museums in Wisconsin and other states. The major collections are covered in Chapter 27. A detailed history of the role of the Milwaukee Formation in the cement industry is covered in Chapter 28.

A

B

Fig. 2. The Milwaukee Cement Company. Both images are courtesy of Historic Photo Collection / Milwaukee Public Library. **A:** Part of the Milwaukee Cement Company operations (circa 1880) located along the Milwaukee River where the Estabrook Park bathing beach would later be developed. **B:** Quarry workers posing for a group photo at #1 Quarry, Milwaukee Cement Company, 1888. Some of these men very likely found many of the fossils shown in this book. Unfortunately, thelr names were not recorded..

Chapter 2 — Stratigraphy

Geology, Paleoenvironment, and Paleobiology

In general the Wisconsin Devonian rocks exhibit a series of three basic and successive, unconformity bound, depositional sequences (Fig. 3) that can be observed throughout much of the adjacent Midwest, including northern Illinois and Indiana, eastern Iowa, and western Michigan. The oldest is an initial transgressive phase consisting of poorly fossiliferous carbonates deposited in marginally marine or hypersaline environments, commonly with evidence of solution breccias. In the Milwaukee area, this sequence is represented by the Thiensville Formation. This sequence is overlain by a series of normal marine rocks of relatively shallow water carbonates and shales represented by the Milwaukee Formation. A final Late Devonian sequence of deeper water shales represented by the Antrim Shale overlies the Milwaukee Formation. Early Devonian rocks are absent from much of this region, with the exception of deeper basins, as the area was generally emergent during this time interval. As a result of this absence, the contact between the underlying Silurian rocks of the Wenlock Epoch with the base of the overlying Middle Devonian rocks in and around Milwaukee represent a major unconformity of 25 or more million years of time.

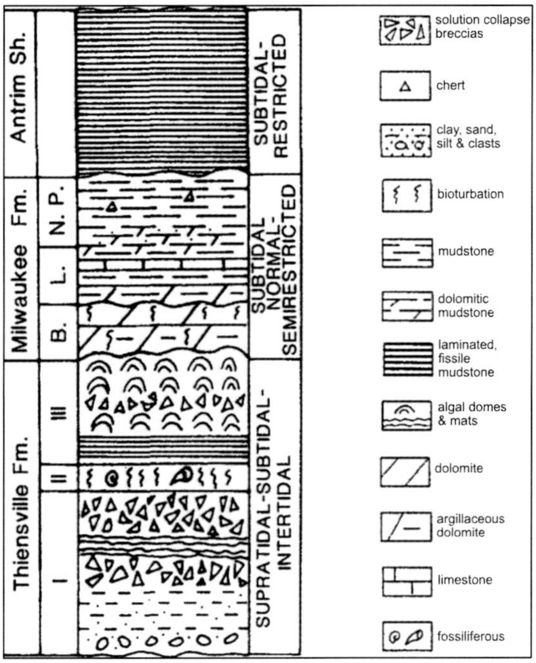

Fig. 3. Stratigraphic column of Devonian rocks in Milwaukee County, Wisconsin, showing dominant lithologies and environmental determinations. I, II, III are informal subdivisions of the Thiensville Formation; B, L, and N.P. correspond to Berthelet, Lindwurm, and North Point Members, respectively, of the Milwaukee Formation. From Kluessendorf, Mikulic, & Carman, 1988, Fig. 4. Copyright © 2009 by the Canadian Society of Petroleum Geologists. Reprinted by permission of CSPG whose permission is required for further use. Symbol legend re-added for clarity.

<u>Thiensville Formation</u>

The basal Middle Devonian rocks of the Milwaukee area are represented by the Thiensville Formation. This unit was deposited during a marine transgressive flooding event covering much of the Midwest. The depositional environments of the Thiensville are supratidal–subtidal–intertidal in nature, indicating deposition in very shallow water in an arid coastal sabkha setting (Kluessendorf, Mikulic & Carmen, 1988); Mikulic & Kluessendorf, 1988a; Mikulic & Kluessendorf, 1998). With both its upper and basal contacts being unconformable, the Thiensville underlies the Milwaukee Formation and overlies either the Silurian Waubakee Dolomite or reefs in the Racine Dolomite (Mikulic,1979; Mikulic & Kluessendorf, 1988a,b; Mikulic

& Kluessendorf, 1998; Kluessendorf & Mikulic, 1998). Historically the quarries of the Milwaukee Cement Company bottomed out at the top of the Thiensville, which initially was thought to be part of the "Niagaran" Formation (Chamberlain, 1877, 1878) or Waubakee Dolomite (Alden, 1906, 1918; Cleland, 1911). The Thiensville Formation was named by Raasch (1935) for a road cut in Ozaukee County north of Thiensville. The formation is 17–23m thick, lithologically complex, and generally grades upward from dirty, argillaceous and poorly lithified sediments at the base to clean carbonate which include algal mat laminates or domes at the top. Kluessendorf, Mikulic & Carmen (1988) subdivided the Thiensville into three subunits: a lower *Subunit I* dominated by pedogenic and evaporitic sediments; a middle *Subunit II* consisting of a brown, granular dolomite containing marine fossils topped by a black shale layer; and a lithologically variable upper *Subunit III*, which is commonly characterized by microbial mats. Subunit II is the only fossiliferous portion of the unit, and represents deposition in a semi-restricted, subtidal environment. The biota of the middle beds consists mostly of brachiopods, rugose corals and bivalves, all in the form of internal and external molds. The Thiensville is considered Givetian; however, the contained fossils are not sufficient for precise age determination.

Milwaukee Formation

The Middle Devonian Milwaukee Formation represents the second sequence in this area. It was named by Alden (1906) for the argillaceous dolomite and shale that was quarried extensively by the late nineteenth-century natural cement industry in Milwaukee County. This unit is Middle Devonian (Givetian) (Schumacher, 1971b; Klug, 1995) in age, and is known to occur only in the northeastern portion of Milwaukee County. The most prominent outcrops of this unit occur along the Milwaukee River at Estabrook Park, and on the Lake Michigan shore of Whitefish Bay.

The Milwaukee Formation contains a diverse and prolific biota (Cleland, 1911; Pohl, 1929; Raasch, 1935; Griesemer, 1965), and is the most fossiliferous unit of the Wisconsin Devonian. During the late 1800s and early 1900s, while this formation was being quarried for cement rock, large fossils collections were made by local naturalists such as Thomas A. Greene, Edgar E. Teller, and Charles Monroe (Mikulic, 1983). Gilbert O. Raasch and Joseph G. Emielity later made additional large fossil collections during the twentieth century.

The Milwaukee Formation appears to have been deposited in relatively shallow, subtidal, normal-marine conditions (Kluessendorf, Mikulic & Carmen, 1988, Mikulic & Kluessendorf, 1988a) at and below storm-wave base (Zambito & Day, 2015). It is divided into three formal members (Fig. 3) by Raasch (1935), all of which represent deposition in normal shallow marine seas and are abundantly fossiliferous. It lies disconformibly below the Frasnian (Upper Devonian) portion of the Antrim Shale and above the Givetian Thiensville Formation (Mikulic & Kluessendorf, 1988a).

The Milwaukee Formation was at various times correlated with the Hamilton Group (Hall, 1862; Chamberlin, 1877, 1978; Teller, 1900; Cleland, 1911) and Tully Formation (Cooper et al., 1967) of New York, the Thunder Bay Formation (Cooper et al., 1942; Cooper, 1967) and Potter Farm Formation (Cooper, 1967) of Michigan, and the Rapid and Coralville Members of the Cedar Valley Formation of Iowa (Griesemer, 1965; Schumacher, 1971b). Based on conodonts, Mayer (2001) correlated the upper beds of the Milwaukee Formation with the Little Cedar Formation and the *subterminus* conodont fauna of the Coralville Formation of Iowa. Griesemer (1965) reported that of the 46 brachiopod species identified from the Milwaukee Formation at that time, 12 were also reported from the Cedar Valley Formation, although taxonomic status has changed considerably since that study was conducted. The Cedar Valley has since been elevated to Group and includes the Little Cedar, Coralville, Lithograph City, and Shell Rock Formations. The Rapid is, at the time of writing, a member of the Little Cedar Formation. The Little Cedar and Coralville Formations are both Givetian (Cordie & Witzke, 2017).

North Point Member

The uppermost member (North Point Member) is the youngest, and is known only from subsurface information including tunneling (primarily from the North Avenue Intake Tunnel) and cores from the more recent Milwaukee Metropolitan Sewerage District's Deep Tunnel project. This unit is approximately 7m (23ft.) thick, consists of dark gray mudstone and dolomitic mudstone, and contains thin layers of dolomitic siltstone with chert and silicified fossils, the latter of which may represent storm deposits (Mikulic & Kluessendorf, 1988a). The fauna is dominated by chonetid brachiopods (Chapter 12), large tentaculitids (Chapter 9), and bivalves (Chapter 14), but also contains some gastropods (Chapter 14), lingulid brachiopods, and other shelly fossils. Its limited faunal diversity, paucity of bioturbation, dark color grading upward from a dolomitic siltstone to a silty mudstone, along with the presence of finely disseminated pyrite (Chapter 4) suggest deposition under restricted, dysaerobic, marine conditions (Kluessendorf, Mikulic & Carmen, 1988). The North Point Member's contact with the Lindwurm is transitional; its contact with the overlying Antrim Shale represents a disconformity (Mikulic & Kluessendorf, 1988a; Zambito & Day, 2015).

Lindwurm Member

The Lindwurm Member is very fossiliferous and ranges from 4.6–6.1m in thickness. Only the lower 1–2m (3–7ft.) of the Lindwurm Member are exposed at the time of writing, but its disconformable contact with the underlying Berthelet Member (Raasch, 1935; Kluessendorf, Mikulic & Carmen, 1988; Mayer, 1997) is well exposed (Fig. 4). The lower beds of the unit are argillaceous dolomite and subsequent beds of dolomitic siltstone or mudstone interbedded with biocalcarenite layers which decrease in thickness and abundance upwards. Diverse and abundant marine fossils occur in skeletal packstone–grainstone interbeds, which indicate deposition under normal-marine conditions (Kluessendorf, Mikulic & Carmen, 1988). Shelly fossils are abundant throughout the member, while fishes (Chapter 20), graptolites (Chapter 18), and land plants (Chapter 22) are also present in places. The Lindwurm is the only member of the Milwaukee Formation that contains a diversity and abundance of echinoderms along with more trilobite taxa when compared to the other Milwaukee Formation members. However, it has a significantly lower diversity and abundance of cephalopods and fishes, both of which typify the underlying Berthelet Member. Schumacher (1971b) correlated the upper Berthelet and lower Lindwurm with the Rapid and Coralville Members of the Cedar Vally Formation of Iowa, and Klapper & Barrick (1983) correlated the Berthelet with the Rapid Member. The Lindwurm–North Point contact is gradational, and partially for this reason, Zambito & Day (2015) suggested that stratigraphic revision is warranted. Large collections of fossils of the Lindwurm were collected by Teller, Monroe and all more recent collectors, but the older Greene collection has relatively few specimens from this unit.

Fig. 4. Vegetation grows at the disconformity between the Lindwurm Member and underlying Berthelet Member; type locality of the Milwaukee Formation at Estabrook Park, Milwaukee. Note the asphalt vugs in top layer of Berthelet Member.

Berthelet Member

The Berthelet Member (Fig. 3) is a fossiliferous, gray, argillaceous dolomite that varies from 6–7m (20–21ft.) thick; only the base and the upper two meters or so are well exposed at the time of writing. The upper 2m (7ft.), which were used as the roof for tunnels in the cement mines, is more dolomitic, thick bedded, vugy, and asphaltic. Brachiopods are common throughout most of the unit, whereas cephalopods and a variety of other taxa are more common in the upper 2m, which have been informally referred to as the cephalopod beds; rare terrestrial plant fragments are also present (Chamberlain, 1883; Kluessendorf, Mikulic & Carmen, 1988). Historically, the Berthelet Member has been one of the most important Midwestern sources of Middle Devonian fish fossils. As is typical of most dolomites of the region, and in contrast with the overlying North Point and Lindwurm Members, most of the fossils of calcareous skeletal elements in the Berthelet are found as internal and external molds. The majority of the fish remains of the Milwaukee Formation are from this member. Most of the formation's land plants, all carbonized to some extent, have come from the lower beds of the Berthelet Member. Cephalopods (Chapter 14), bivalves, and asphalt vugs (Chapter 4) are common in the upper beds of this member.

The top of the Berthelet Member is marked by an important irregular burrowed surface, and phosphatic nodules occur in the basal Lindwurm, indicating a depositional hiatus (Kluessendorf et al., 1988). The basal contact is unconformable and mineralized with Thiensville clasts occurring in the basal Berthelet. A similar depositional break occurs between the Dundee Formation and Traverse Group in southeastern Michigan (Kluessendorf, Mikulic & Carmen, 1988). The Berthelet appears to represent a transition from semi-restricted subtidal beds of the Thiensville to more normal-marine subtidal conditions (Kluessendorf, Mikulic & Carmen, 1988).

Antrim Shale

The Late Devonian Antrim Shale represents the uppermost of the three depositional Devonian sequences of the area. The Wisconsin occurrence of the Antrim Shale represents the youngest Paleozoic rocks found in Wisconsin and is known only from the subsurface in a very limited area of the northeastern part of Milwaukee County (Fig. 1). The name "Antrim Shale" was introduced by Schumacher (1971a) as a Michigan-derived replacement for the "Kenwood Shale" which Edwards & Raasch (1921) had named these strata (Klug & Nelson, 1977). These strata were considered to be of Mississippian age (Pohl, 1929; Raasch, 1935). Based on more recent conodont studies (Schumacher, 1971a; Klug, 1995; Mayer et al., 1998), the Antrim is now known to be Late Devonian (Frasnian–Famennian) in age and correlates with the Antrim in the Michigan Basin to the east. Based on conodonts (Chapter 19) recovered from the lower 1m of this formation, Mayer (2001) correlated those strata with the Frasnian Zone 4, in the uppermost part of the Lithograph City Formation of Iowa. The Antrim is dominantly a finely laminated, organic-rich, fissile, black mudstone with thin coal seams (Raasch, 1935). Fossils from the Antrim Shale in Milwaukee include lingulid and productid brachiopods, bivalves, gastropods, scolecodonts (Chapter 15), and ptyctodontid plates and spines (Raasch, 1935; Mayer, 2001). The sediment and limited faunal diversity suggest deposition under restricted, dysaerobic to anoxic conditions (Kluessendorf, Mikulic & Carmen, 1988). The Antrim overlies the Milwaukee Formation unconformably and is, in turn, overlain unconformably by Quaternary sediments representing a gap in time of over 350 million years.

Chapter 3 — Life in Devonian Milwaukee

The Milwaukee Formation was deposited, and its contained biota lived, during the Middle Devonian, about 385–383 million years ago. If we could go back to that place and time, we would see assemblages of animals and plants that at first glance appeared not greatly different from the life in modern seas (see frontispiece and front cover). If we didn't look too closely, we might even think we were diving into the shallow waters in one of today's oceans. Among the diverse life would be colorful communities of echinoderms adhering to the substrate (Chapter 17). In other parts of the seafloor would be brachiopod-rich communities of these two-valved animals living attached or free-lying on the seafloor and filtering sea water for suspended plankton using tentacle bearing lophophores. Thickets of bryozoans (Chapter 10) or twig-like or fan-shaped bryozoans also filtered sea water for plankton. In still other areas would be mollusk-dominated communities consisting of an assortment of bivalves (clams and their relatives), gastropods (snails and their relatives), and related groups (Chapter 14). Between and overlapping some of these communities, we would see a profusion of arthropods (Chapter 16) and other invertebrates living on the seafloor. Traces of their movements on and within the substrate would be made (Chapter 23). Swimming around and above us is an assortment of fishes of all sizes (Chapter 20)—some very large and menacing—as a variety of plankton drifts slowly by. Occasionally, we would observe a piece of land vegetation floating into view and sinking to the bottom. Although this scenario describes a scene that applies to life in Devonian seas, it applies to the waters of today's shallow seas as well. But huge differences would soon strike any discerning eye.

Paleoenvironments

During deposition of the Milwaukee Formation the region of Wisconsin was adjacent to the Michigan Basin—a deeper water, gently subsiding area to the east and northeast. Water shallowed to the southwest into the area of Iowa where the Middle Devonian is represented by tidal flats and lagoons or during times of relatively high sea level shallow shelf settings. The climate was subtropical to tropical as this region lay near the equator.

Water Depth

Some indication of water depth is especially important in any study of ancient marine environments, because depth is correlated with many other variables (e.g., temperature, light, and environmental energy) that have a strong controlling influence on benthic (bottom dwelling) organisms. Unfortunately, depth is also a difficult parameter to assess, as there is little direct information on water depth. The Milwaukee Formation shows some minor evidence of deeper storm wave-current deposited sediments. These include thin (1–10 cm) wavy beds that show micrograding, and accumulation and deposition from gradient currents and/or combined flows that involve minor wave action. These deposits of silt-sized carbonate sediment were swept into the muddy basin during times of strong backflow currents or gradient currents. These silty limestone beds were probably deposited toward the deeper end of storm wave-current effects, i.e., at deep storm wave base. For shallow interior seas, this position may have been as much as 50m (~150 feet) or slightly more but clearly the seafloor was not in very great depths. Very gentle distal ends of gradient currents may have shifted sediment and shells on the seabed from time to time. However, the currents were not strong enough to impart a very strong preferred orientation. In shallower depths, storm waves locally broke up colonies of bryozoans and shells and reworked them into local lenses of slightly broken debris.

Shallower water sediments are indicated in some of the correlative strata in Iowa in the Coralville Formation which are composed of crinoidal sand with abundant thickets and patch reefs of corals and sponges called *stromatoporoids*, representing probably no more than 10–15m of water. These beds also contain locally abundant colonial corals and stromatoporoid sponges that probably required shallow, moderately well-lit waters. Milwaukee Formation sediments lacked reef development and many of its strata were deposited in considerably deeper water. It is also possible that the sediments of the Milwaukee Formation were too clay rich, were either episodically or continuously deposited too rapidly, or that the substrate was too poorly lithified to support coral or stromatoporoids growth.

Light Intensity

Evidence for ancient light levels is tentative, needing more detailed study. The presence of relatively sophisticated *schizochroal eyes* in the abundant trilobite *Eldredgeops* points to the presence of some light, as organisms living in perpetually dark settings (such as cave crayfish) tend to have weak or non-existent eyes. Eyes of that trilobite and other phacopids were adapted for forming clear images and this requires sufficient light. The large size of the eyes, on the other hand, may indicate relatively low light levels to which the organisms were adapted. This seafloor was undoubtedly muddy and prone to high *turbidity*; this indicates depths of just a few tens of meters. Together these lines of evidence suggest deposition in the *dysphotic* to lower *euphotic* zone, at depths of perhaps 30–60m of water, deeper to the north where facies were mainly dark, barren shales, reflecting stagnant, dysoxic environments.

Salinity

Salinity in the Milwaukee Formation depositional basin was mostly normal marine (about 35 parts per thousand salinity) as evidenced by the numerous organism groups that are known to be quite intolerant to conditions other than normal marine, such as brachiopods, echinoderms and cephalopod mollusks.

Substrates and Sedimentation

The substrates of the Milwaukee Formation were mainly soft muds but with local accumulations of skeletal hard parts. It is clear that the distribution of certain organisms was more specifically controlled by substrate. Many of the brachiopods, especially the chonetids and *Mucrospirifer* were evidently adapted to a "snowshoe" mode of life suggestive of adaptation to soft, fluid muds (see below). Likewise, the broad, flattened shape of some bryozoan colonies, such as *Lichenalia*, may have been adaptive for distributing weight on muddy seafloors to prevent sinking into the substrate.

Other brachiopods (attached by pedicle stalks), bryozoans, corals and a majority of echinoderms required hard substrates for settlement. In some areas stable, hard substrate was provided by skeletal debris, particularly brachiopod shells. These shelly organisms provided further substrates for other encrusters such as bryozoans, hederelloids (Chapter 11), and edrioasteroids (Chapter 17).

The process of "taphonomic feedback" (Kidwell & Jablonski, 1983) no doubt played a critical role in the buildup of local patches of more diverse organisms. This means that the accumulation of dead shells, initially of "pioneering" mud dwelling brachiopods (like chonetids), bivalves, and trilobite molts, provided substrates for more diverse communities, including twig bryozoans and echinoderms, which, in turn, may have been favored by intervals of lower muddy sediment influx.

Environmental Energy

Milwaukee Formation seafloors were predominantly low energy (current velocities generally not exceeding 20cm/sec). Low current energy means that organisms could be relatively weakly attached to the seafloor and still maintain their position. Some organisms were free lying or weakly tethered by small holdfasts (crinoids, probably blastoids), cemented pads (graptolites and bryozoans) or pedicle stalks (brachiopods). Evidently, this was adequate anchorage for stability during the majority of the time. However, during the rare storm events such weak tethering became a major weakness as storm waves and currents disrupted the normally quiet seafloor, uprooting millions of bryozoans, echinoderms, brachiopods, and other organisms and in some instances rapidly burying them in layers of muddy sediment resuspended by waves from shallower areas.

The normally low current and wave energy settings of deeper Milwaukee Formation environments may have limited certain organisms that required water movement to provide feeding currents, such as with most echinoderms. It may also have promoted a multi-tiered community in which different groups of organisms fed at different levels above the seafloor (Ausich & Bottjer, 1982). Low-level organisms occupy a zone of abundant suspended food particles just 1–2cm above the seafloor. Intermediate level *suspension feeders* such as many bryozoans, and some crinoids are elevated about 5–10cm above the bottom, while some long-stemmed crinoids may feed at levels exceeding a meter above the seabed. Brachiopods and bryozoans, which could produce their own filter feeding currents (active filter feeders) may have benefitted from living close to the seafloor where suspended food was concentrated by gravity settling, whereas crinoids that relied strictly on external currents (passive suspension feeders) survived by elevating their crowns up away from the stagnant boundary zone near the sediment surface and into levels of at least some current energy.

Modes of Life of Devonian Organisms and Community Paleoecology

Marine ecologists classify organisms into *guilds* or generalized modes of life based upon their: a) position with respect to their life on the seafloor (*benthic*), or in the water column, either free floating (*planktonic*), or swimming (*nektonic*), as well as b) feeding or trophic group. Benthic organisms are further identified as *infaunal* (living in burrows within the sediment) or *epifaunal* (living on the sediment surface). Certain epifaunal organisms, like trilobites and gastropods, were vagrant, crawling on the seafloor. But many more, such as corals, brachiopods, bryozoans, and crinoids, are sessile or sedentary. Ever-present in modern and Devonian seas are the epibionts that attach to or encrust living surfaces and sclerobionts that attach to or encrust non-living surfaces; however, the content of the two biotas stands in sharp contrast to one another. Today a number of organisms encrust hard objects on the seafloor, including on other organisms. These encrusters incude oysters, barnacles, and spirorbid worms, none of which were present in Devonian Milwaukee; however, the fossils of the Milwaukee Formation demonstrate that many other animals made sure that those surfaces were fully exploited, as will be discussed below.

All ecosystems have complex interactive webs or trophic pyramids that require an external energy source (generally sunlight), including organisms such as green plants and algae that produce their own food from simple inorganic substances such as carbon dioxide (*autotrophs*), animals that feed on plants or other animals (*heterotrophs*), and *decomposers*, primarily bacteria, that degrade organic matter and recycle its contained nutrients. Such relationships based upon modes of feeding are called *trophic systems*. In the case of marine systems, common trophic modes include removal of suspended planktonic algae and zooplankton or particulate organic detritus from seawater. Such *suspension feeding* may include passive suspension feeding, as in crinoids, which rely upon external currents in seawater to bring food particles to sticky feeding surfaces, and active *filter feeding* in which organisms create their own feeding currents using movements of flagella (as in sponges), cilia (e.g. in brachiopods and bryozoans) or whole limbs, as in certain arthropods. Infaunal organisms, termed *deposit feeders*, live in the substrate and actively ingest sediment and extract food from it. In addition, other organisms are *scavengers* or *grazers*, rasping or scraping up dead or living organic matter, typically plants. Others are *predators*, mainly active or vagrant swimming or crawling forms that kill and consume other animals. The Milwaukee Formation biota included all of these trophic systems.

Interpretation of the modes of life (*paleoautecology*) and ecology of ancient ecosystems (*paleoecosystems*) involve a combination of knowledge of the modes of life of the closest living analogs, and of functional morphology, which uses the form of skeletons to infer their function, in conjunction with direct evidence from fossils and surrounding sediments and information on the biology of mostly closely related living organisms. Certain animal groups from the Milwaukee Formation serve well to illustrate some of these modes of life and paleoecosystems.

<u>Brachiopod Life Modes</u>

Brachiopods are bivalved organisms, unrelated to mollusks, that still live today much as they first did more than a half billion years ago, although they are far less numerous and varied than in the past; during the Devonian they were typically the dominant shellfish in the sea. All were low level sedentary filter feeders. There are three basic groups of living and fossil brachiopods: the *lingulids* bear a phosphatic, fingernail shaped shell lacking hingement or articulates, *craniiforme* brachiopods have small circular calcitic shells, which typically encrust other shells, and *rhynchonellforme* brachiopods, formerly called articulate brachiopods because of their interlocking hinge teeth at which the calcitic valves pivot upon each other.

The somewhat more primitive lingulids are undoubtedly among the most conservative animals on the planet having survived nearly unchanged since Cambrian times. They are capable of locomotion using an elongate muscular stalk or pedicle to burrow in sediments. Modern lingulids live infaunally within sediments in burrows in tidal flats as well as offshore settings and are tolerant of a range of conditions including mixed and low salinities and rather low oxygen conditions.

The craniiforme brachiopods are nearly as ancient as lingulids and throughout their history have tended to cement to hard substrates by a scab-like pedicle valve and produce a gently domed brachial valve. These brachiopods are found in some number in the Milwaukee Formation where they are found encrusting other shells rather like the unrelated oysters.

Today rhynchonelliform brachiopods are represented by small bilaterally symmetrical globose shells sometimes called “lamp shells” that tend to live on rocky seafloors and reefs where they tether their shells by tough fibrous pedicle stalks. But in the Devonian, brachiopods had much more varied sizes and shapes reflecting more varied modes of life.

Brachiopods generate feeding currents using ciliated tentacles on their looped or coiled *lophophores* that produce rhythmic beats that propel water and suspended particles. In many brachiopods, the surface area for collecting suspended food was greatly increased by the development of strongly coiled lophophores supported by spiral ribbons of calcite (*brachidia*) which may actually be preserved in the fossil record. These spiralia evolved separately in the spiriferids (which are named for their spiral brachidia) and in the atrypids. Many of these organisms have morphological strategies that may have aided in water flow. For example, the groove at the center of the valves (fold and sulcus) in *Mucrospirifer*, *Orthospirifer*, and other spiriferids, as well as rhynchonellids such as *Cupulorostrum* and *Eumetabolotoechia*, served to funnel the wastewater away from the lateral incurrent streams and may have had some Venturi effect that helped to promote water flow.

Many brachiopods may have initially settled on small hard substrate objects. Some spiriferids, such as *Orthospirifer* and *Cyrtina*, may have attached by pedicle stalks but also developed broad triangular shaped interareas around the pedicle opening. These broad surfaces probably also spread the weight of the animal over larger areas to stabilize the shell on soft substrates. The atrypids and rhynchonellids (e.g. *Cupulorostrum*) were attached by pedicle stalks and may have initially settled on shell fragments and other skeletal materials. The robust ribs and zig-zag *commissures* (openings of the valves) may have strengthened the shell against physical damage or predation.

Other forms with broad, relatively flat or concavo-convex shells, such as the strophomenids were free-resting on the seafloor as adults. The large, broad surface area of shells in forms like *Protoleptostrophia* and *Strophodonta* may have helped to spread out the weight of the shell of these free-resting brachiopods, like a snowshoe, to prevent them sinking into soft muds. Other common brachiopods such as *Mucrospirifer* and *Tylothyris* employed a somewhat different strategy of greatly elongating the shell along the hingeline. In these brachiopods, the corners of the shell were even extended laterally, providing outriggers that probably also served to support the shell at the substrate like a pair of cross country skis on soft snow. The small chonetids appear to have been opportunistic forms that thrived in somewhat stressed, low energy. Somewhat like *Mucrospirifer*, they possessed a series of spines directed outward from the hinge that no doubt served as anchorage on soft, fluid muds. In addition, their small size and very tiny internal volume meant that they put minimal weight on the soft substrate and had minimal needs for oxygen and suspended food. This means that these brachiopods could colonize relatively muddy, low oxygen areas of the seafloor. Even more tolerant were the small globose brachiopods *Eumetabolotoechia*; these thin shelled forms often lived in the deepest and lowest oxygen seafloor settings where most other organisms were excluded.

Bryozoan Life Modes

Bryozoans or "moss animals" go largely unnoticed in modern marine environments but they are abundant and highly varied, especially in subtropical reefs. Bryozoans are colonial organisms which construct encrusting, sheet-like, shrubby, stony, or lacy colonies of calcite. The feeding individuals are tiny polyp-like *zooids* which, like their distant relatives, the brachiopods, filter seawater for suspended plankton, using feeding currents generated by their ciliated tentacles of their lophophores.

Most bryozoans are sensitive to high turbidity, probably because suspended silt may clog their delicate filtration meshes. However, certain Devonian forms, such as the stick-like *Sulcoretepora,* branching trepostomes, and the lacey funnel-shaped fenestrates, were capable of surviving in relatively muddy seafloor environments not only because of active filter pumping but also elevation of growing portions of their colonies into areas above the low energy seafloor. Their slender branching shapes, adaptive for low energy environments, may also have helped in shedding sediment from the surfaces of colonies. Other bryozoans and the related hederelloids formed encrusting patches on other substrates such as brachiopods. This provided a clean, non-turbid area for the delicate filter feeding organism.

Echinoderm Life Modes

Many adult echinoderms living today are mobile, yet most echinoderms found in the Milwaukee Formation were *sessile*—attached to other objects for most or all of their adult lives. Some of this distinction is situational, as fossils of the very mobile starfish (asteroids), sea urchins and sand dollars (echinoids), and sea cucumbers (holothurians) are absent from the Milwaukee Formation, but were sometimes preserved in Devonian and much older rocks of other areas. It may be that those animals were also among the fauna in Devonian Milwaukee—the absence of their fossils being due to the difficulty in preserving them rather than an absence of the animals. The majority of the adult crinoids in today's oceans are free-swimming or crawling, yet most of those from the Milwaukee Formation and other Devonian rock units appear to have been sessile, living their adult lives adhering or attached to the substrate by the *holdfasts* of their long stalks. The now-extinct blastoids lived alongside the crinoids in Milwaukee's Devonian sea. Like those crinoids, they were stalked and sessile. Edrioasteroids, the early cousins of the sea urchins and sand dollars, are also now extinct but were present in Wisconsin's Devonian sea attached to brachiopods (Chapter 17).

Echinoderms do not generate their own feeding currents; therefore, many, such as the edrioasteroids and stalked crinoids and blastoids, were passive *suspension feeders*, dependent upon external currents to bring suspended food particles into the vicinity of their food grooves, which were covered with *tube feet* and mucous to trap these particles. By analogy with living stalked crinoids, many Paleozoic forms with long, multi-branched or even compound *pinnulate* arms, such as *Melocrinites*, are interpreted as having formed filtration fans. Modern crinoids are "leeside" suspension feeders; that is, they recurve the backsides of their arms into the strongest water currents. Water flushes through the spaces between pinnules and forms a low-pressure zone on the other side where the *ambulacral grooves* are amply supplied with sticky tube feet. Particles spiraling in these eddies are captured on the tube feet and propelled into the cilia lined food groove; a combination of food and mucous then makes its way to the centrally located mouth like a series of stream drainage systems.

Densely pinnulate arms were efficient for capturing suspended food, but only if sufficient currents existed in the environment to force water through the relatively dense meshwork of arms and pinnules. Although modern crinoids have yielded insight into crinoid feeding behavior, it is evident that not all Paleozoic crinoids could have formed efficient parabolic feeding fans. Flexible crinoids like *Taxocrinus* (front cover) had broad, space-filling brachials in the arms that ended in tendril like branches, but lacked pinnules. Their arms are often preserved as tightly coiled, like a clenched fist. It has been suggested that these crinoids were raptorial, actively capturing larger plankton with their flexible arms and pulling it into the mouth.

The blastoids lacked arms and instead had tiny, thread-like *brachioles*. These were so slender that there was little or no space for tube feet. It has been suggested that these organisms lacked tube feet altogether (Sprinkle, 1973). If so, they must have fed in a very different way than crinoids, perhaps using mucous secretion and cilia to trap microplankton. The small size of their food collecting appendages may have confined most of these organisms to moderate energy or more food-rich shallower environments. However, the small size and specialized *hydrospires* for respiration may have helped a few forms like *Hyperoblastus* survive in somewhat low oxygen, muddy offshore settings.

Mollusk Life Modes

One of the most conspicuous mollusks in today's seas, the oyster, was completely absent from the Milwaukee Formation and other Devonian occurrences. Oysters did not evolve until later in the Mesozoic Era. Nonetheless, many of the other groups of bivalves, such as clams, scallops, and mussels, were fairly well represented, and did not look much different from some modern forms.

These include the ubiquitous *Palaeoneilo* and other nuculid bivalves. These small "nut clams" are known from modern seas and are a prime example of a living fossil group which has survived with little modification—at least of the shell—since early Paleozoic times. These bivalves are known to be very active, shallow sediment burrowers that use a muscular foot to rummage through muddy sediments and extracted food particles using a pair of tentacles known as *labial palps*.

Other bivalves lived infaunally or semi-infaunally within the muddy sediments being anchored to particles in the mud, such as shell fragments, using tough elastic threads of *byssus* which they secreted. These clams like modiomorphids, *Grammysia* and *Cimitaria* were filter feeders that brought currents into their mantle cavities using cilia and filtered the water with intricate gills which served both for respiration and food capture. Unlike many modern bivalves these Devonian forms lacked *siphons* which are tubes that aid in sucking in currents and separating these from excurrents.

Gastropods or snails were also common. One conspicuous difference was the relative scarcity of thick-shelled, highly ornamented, and large gastropods, as compared to the common occurrence of those mollusks in today's seas. Gastropods had varied modes of life; most were active organisms that utilized a muscular foot to crawl on the seabed and a rasping tongue or *radula* for feeding. Some of the more typical platyceratids in the Milwaukee assemblages, such as *Cyclonema* and *Naticonema*, may have lived as active benthic scavengers or predators. However, the majority of this group was adapted to an unusual mode of life. These organisms attached, probably as juveniles to the *tegmens* of certain crinoids, especially *Melocrinites,* and took up a dependent, sedentary life fixed to their hosts. These gastropods were positioned over the anal vents of the echinoderms where they apparently fed upon fecal material expelled by their hosts; that is, they were *coprophagous*. They have generally been interpreted as *commensals* that benefitted from their hosts but did them little or no harm. However, some authors have argued that these snails could also be parasitic. By the Devonian there is evidence that some platyceratids used their radulas to drill into the gut of their hosts and rob them of some food (Baumiller & Gahn, 2002). Some morphologically similar Silurian camerate crinoids and the rhombiferan cystoid, *Caryocrinites*, have been found with attached platyceratids. The association of these related and similar appearing crinoids, along with a prominent cystoid, all with attached platyceratids, suggest that morphologic and behavioral adaptations associated with rheophilic feeding were the primary factors controlling this association (Kluessendorf, 1983). It is possible that some of these commensal platyceratid snails could have become partially or totally parasitic by the later Paleozoic, but association of a specific echinoderm morphology and feeding behavior with platyceratids is a more compelling factor in recognizing these relationships as coprophagous behavior. A single specimen of *Melocrinites* from the Lindwurm Member of the Milwaukee Formation has been found with a commensal platyceratid attached.

Arthropod Life Modes

Due to the lack of mineralized skeletons in many arthropods, that phylum is, except for trilobites, not as well represented in the fossil record as are such animals as gastropods, bivalves, brachiopods, and bryozoans. We do know, however, that the Milwaukee Formation does contain several species of trilobites, as well as phyllocarid and ostracod crustaceans. Trilobites were already on their way out by the Devonian and were extinct by the end of the Paleozoic Era, while crustaceans continued to diversify, and the arthropods as a whole went on to become the most species-rich phylum on the planet.

Trilobites may have been partly *detritivores* or scavengers feeding on decaying organisms; they lacked biting mouthparts and so were constrained to feeding on relatively soft food. Certain trilobites, however, were probably predators on soft-bodied organisms. By analogy with modern horseshoe crabs, trilobites may have utilized the bristly inner joints of their legs (gnathobases) to grasp soft bodied organisms such as worms from within the sediment and pass them forward to the mouth, which was underlain by a wedge-shaped plate, the *hypostome*. In some cases, trilobites may have worked the food between anterior legs and the hypostome to break it down into small pieces to be ingested by a suctorial mouth. Most trilobite fossils are represented by isolated cephala, cranidia, free cheeks thoracic segments, or hypostoma. This disarticulated nature of trilobite fossils generally results from the trilobite periodically molting during growth, although articulated thoraxes and pygidia may also be molted exoskeletons, while isolated parts might also result from decay of dead individuals.

Infaunal Life Modes

Living within the sediments beneath bodies of water, modern infaunal organisms include certain species of arthropods, mollusks, holothurians, chordates, and many phyla of worms, including annelids. Most likely, species of those groups also lived infaunally in Devonian seas. Infaunal deposit feeders on the Middle Devonian seafloor are represented primarily by trace fossils, particularly the feeding burrows *Planolites* and *Chondrites* that are abundant and/or well preserved in silty beds. These are traces of sediment mining "worms" that do not clearly indicate their originators. As noted, the small "protobranch" bivalves like *Palaeoneilo* were also active deposit feeders and trilobites may have burrowed for food; some deposits show distinctive traces that are attributed to trilobites feeding in the sediment. Additionally, it is very likely as noted above that some of the linguliform brachiopods of the Milwaukee Formation were infaunal, as are living survivors of that brachiopod group.

Nektonic Life Modes

Nekton consists of the animals that are able to swim against the currents in the open waters, and the majority of them are (or were, in the case of the Devonian forms) one kind of fish or another. But that is essentially where the similarity ends. Most of today's fishes are teleosts, which did not even appear until the Triassic Period. In sharp contrast, the dominating fossil fishes from the Milwaukee Formation were members of the long-extinct Placodermi. Sharks (part of the chondrichthyes) and lobe-finned fishes (sarcopterygians) swam in Milwaukee's Devonian sea, as well, as indicated by fossils of their fin spines and teeth present in the rocks. Another extinct form of nekton during that time were the conodonts—the tiny eel-like protovertebrates whose microscopic, tooth-like phosphatic fossils, also termed conodonts, are found in the Milwaukee Formation.

The master predators of the Devonian sea were undoubtedly the jawed fishes, but nautiloids and other cephalopods also lived a predatory lifestyle. Cephalopods from the Milwaukee Formation were also conspicuously different from those of today. Whereas the former includes nautiloids, actinoceratoids, and ammonoids, today's cephalopods are dominated by coleoids, which include squids, octopuses, and cuttlefish. Restricted to the oceans' deeper waters, only a few nautiloid species survive in today's oceans. Ammonoids became extremely diverse in the Mesozoic Era and met an abrupt extinction at its end.

Based on modern analogs, notably the chambered *Nautilus*, the nautiloids and ammonoids were streamlined, actively swimming predators. By analogy to *Nautilus*, they probably swam using a form of jet propulsion by taking water into their mantle cavity and forcing it out through a narrow nozzle referred to as a hyponome. Relatively sophisticated eyes enabled nautiloids to track prey. Nautiloids had a cluster of tentacles for seizing other organisms such as trilobites. These surrounded a mouth equipped with a sharp, parrot-like beak that could tear into their prey.

Planktonic Life Modes

The medusoid jellyfish (especially scyphozoans) and the comb jellies (ctenophores) are not represented in the Milwaukee Formation, but they are known to have been alive during that time. As with so many other animals, their lack of hard parts accounts for their scarcity in the fossil record. A group of colonial, net-like animals found in the Milwaukee Formation called dendroid graptolites was sessile but may have been capable of attaching to floating objects as well, thus benefiting from a planktonic lifestyle. A similar lifestyle was possible for other organisms, such as some of the solitary, pyramidal, polypoid scyphozoans known as conulariids, which have been found attached to other objects at their apices (Van Iten et al., 2016b). Neither the graptolites nor conulariids survived into the Cenozoic.

Terrestrial Life Modes

These are represented in the Milwaukee Formation by the land plants and fungi that were washed into the sea and sank to the bottom. Their fossils consist of fragments of lycopods, possible cladoxylopsids and giant, pole-shaped fungi. These organisms, typical of those from the contemporary terrestrial environments of other regions, were tree-like in size, and were thus quite different from the lycopods and fungi of today. Flowering plants had not yet evolved.

Organism Interactions

As in modern communities, organisms on Devonian seafloors were involved in a variety of interactions with other species. While many organisms may simply live side by side on the sea bottom without really interacting, a form of neutralism, others are involved in co-action pairs that include competition, predation, and symbiosis.

Symbiosis is a relationship between individuals of two species of organisms in which at least one species benefits; the relationship may benefit, harm or have no effect on the other species. Wonderful examples of symbiosis abound in the Milwaukee Formation's fossils. They are in fact so common that it is sometimes difficult to carefully examine a macrofossil without seeing evidence of organisms of one kind or another attached to or having been attached to it (Fig. 5).

Many of these biological relationships of two different species are likely to have represented *commensalism*, in which the epibiont benefits by attaching itself to, but does not benefit or adversely affect the host species. With fossils it is more difficult to determine the effect the commensal had on the host. As noted, certain crinoids (including some of those of the Milwaukee Formation) are known to have served as hosts to platyceratid gastropods, which presumably fed on the echinoderm's feces; however, it is generally not certain whether the echinoderm was helped, harmed, or unaffected by this relationship.

As with modern examples, the typical host species was relatively broad and could be either sessile or vagile. The most common hosts in the Milwaukee Formation are cephalopods, brachiopods, and bivalves, while massive bryozoans, echinoderms, and other organisms played host as well. The epibionts ranged from agglutinated foraminifers (Chapter 5), to reptant or encrusting bryozoans, to brachiopods, tabulate corals (Chapter 8), microconchids (Chapter 9), edrioasteroids, and other sessile animals that needed a place on which to attach. Some commensals were picky about the host that they chose. For example, the Milwaukee Formation's edrioasteroids have thus far only been found on atrypid brachiopods.

Fig. 5. Five examples of symbiosis; **A:** MPM P28755 A-B; the brachiopod *Strophodonta halli* Cleland, 1911, hosting the tabulate coral *Aulopora* cf. *A. annectens* Clarke, 1885; scale bar 1cm. **B:** MPM 15692; A brachiopod hosting the bryozoan *Hederella cirrhosa* Hall, 1881; scale bar 2mm. **C:** A splayed bivalve (*Grammysia*, UWM-GM 56200) hosting two linguliform brachiopods (Orbiculoidea?); scale bar 1cm. **D:** An atrypid brachiopod hosting an edrioasteroid (possibly *Krama* sp.), MPM 15471; scale bar 5mm. **E:** A cephalopod hosting several specimens of the microconchid *Microconchus* sp.; UWM-GM G5467; scale bar 5mm.

Determining whether the host was alive or dead when the epibiont attached itself can be difficult or impossible to establish for species that are known only from their fossils. Little to no wear, breakage, or decomposition of the host indicates a greater chance that the host was alive, but does not provide definitive proof. Better evidence includes cases in which the epibiont, such as a bryozoan or tabulate, is encrusted around an entire shell, as occurs with many cephalopods from the Milwaukee Formation (Cleland, 1911; Hilderbrand & Zambito, 2015). The curious *Reptaria* formed a colony of tubes that tended to encrust a living nautiloid cephalopod with its apertures all directed forward and was a hitchhiker on the moving host, growing parallel to the long axis of its host to take advantage of the water flow over the jet-propelled shell (Baird et al., 1990).

Competition is difficult to demonstrate in modern communities and nearly impossible in ancient paleocommunities, yet it may be widespread and possibly an important driver of evolution. In general, organisms tend to avoid competition as it wastes energy that can otherwise be put into growth and reproduction. This may drive diversification of different ways of making a living to avoid overlap—so-called niche partitioning. Some have argued that tiering, discussed above, may be a form of niche partitioning among epifaunal suspension feeders, which avoid duplication by feeding at different levels above the seafloor. There are possible cases recorded in which for example two bryozoans compete for space and one overgrows the other.

Predation, in which one organism kills and consumes another, is also difficult to prove in the fossil record, but there are a number of lines of evidence. In some cases, such as the great shearing jaws of placoderms or the sharp slicing teeth of sharks, there is strong evidence for predation based on functional morphology (Chapter 20). In other cases, as with cephalopods, the argument is more indirect and based mainly on comparison with living organisms. Trace fossils can also provide more direct evidence for predation or failed predation. Such evidence includes gastropod drill holes in clams from the Milwaukee Formation (Chapter 14) and bite marks found in placoderm dermal plates from that formation (Chapter 20) and in contemporaneous clams from New York state (Nagel-Meyers et al., 2011).

Predation is also an important evolutionary driver. An interesting potential example is the rise of spinosity in many invertebrates in the Middle Devonian as a possible anti-predatory strategy in potential prey organisms (Signor & Brett, 1984). Crinoids provide an instructive example. Prior to the Devonian Period, almost no crinoids possessed spines. However, in the Middle Devonian and increasingly into the Carboniferous, a significant proportion of crinoids (up to a quarter) developed spines on their calyxes or arms (e.g., *Melocrinites*, Chapter 17). Signor & Brett (1984) showed a strong correlation of this increased spinosity with the rise of hard shell biting and crushing predators, particularly the jawed fishes, during the Devonian. It is especially correlated with those crinoids that hosted platyceratid gastropods. In the 40 million years prior to that time, while platyceratids inhabited crinoid hosts, none of the crinoids were spiny. But a majority of Middle Devonian host crinoid species, including *Melocrinites*, developed significant spines. Brett (2003) argued that this correlation indicates targeting. Even if crinoids were not particularly tasty (as is seemingly true of modern forms), the presence of attached "escargot" may have caused these hosts to be frequently attacked. Accordingly, they may have responded by preferentially developing spines.

Chapter 4 — Taphonomy

The Milwaukee Formation exhibits a wide range of fossilization modes that help us recognize the rich diversity of its fossil biota. Together, these modes are capable of preserving not only hard parts, but some unmineralized ones and traces as well. Additionally, depending on the mode of fossilization, features of both the outside and inside surfaces of shells can be preserved in detail.

The life in Milwaukee's Devonian seas was undoubtedly subject to the same kinds of post-mortem processes encountered by organisms in today's shallow, warm seas. These processes begin after death of organisms or after parts of skeletons are discarded, such as the molted exoskeletons of arthropods. Soft parts are rapidly degraded and lost by bacterial and chemical decomposition, or serve as a meal for scavengers. Skeletal hard parts are also subject to breakdown by processes that include disarticulation, fracturing, abrasion, and chemical dissolution. The effects of these processes vary somewhat with temperature, pH, salinity, currents, and other factors, but these are significantly reduced once the carcass becomes buried. For optimal preservation, it is therefore important that burial takes place as soon as possible after death.

While the rate of burial and other factors in play during deposition of the Milwaukee Formation were in most cases not sufficient to preserve soft tissues, conditions were often favorable for fossilization of shells, bones, plant remains, and other mineralized and unmineralized skeletal materials.

Skeletal Minerals and Preservation

The dominant minerals or organic materials in the skeletal remains of most of the biota that frequently fossilize in the Milwaukee Formation can be classified into the following categories.

Calcite and Aragonite

The most common polymorph of calcium carbonate, calcite is the main constituent of the hard parts of echinoderms, corals, bryozoans, craniiform and rhynchonelliform brachiopods, microconchids, and trilobites that abound in the Devonian Milwaukee (Fig. 6) and other Paleozoic units. The resulting permineralized fossil is relatively hard and durable as well, regardless of what mineral crystallized within its pores. The skeletal pieces of echinoderms including the Milwaukee Formation's crinoids, blastoids and edrioasteroids, consist of porous single crystals of calcite.

The calcareous shells of most organisms also include aragonite, which is another polymorph of calcium carbonate; these include most mollusks and modern corals. Due to its unstable nature, few Devonian fossils contain any aragonite. Most of the Devonian aragonite has transformed to calcite or dissolved. Aragonitic skeletons tend to undergo early dissolution and to be preserved as molds.

Fig. 6 (top, opposite). Calcareous skeletal materials; Lindwurm Member. **A:** The trilobite *Eldredgeops? norwoodensis* (Stumm, 1953); WESM-KCG 1126; scale bar 5mm. **B:** The bivalve *Palaeoneilo persinnuata* Pohl, 1929; MPM 13757; scale bar 5mm. **C:** Calyx of the blastoid *Pentremitidea* sp.; WESM-KCG 890; scale bar 2mm. **D:** The brachiopod *Orthospirifer clelandi* (Griesemer, 1965); MPM 16580; scale bar 1cm. **E:** The bryozoan *Eridotrypa appressa* (Ulrich, 1890); UWM-GM 5880; scale bar 5mm. **F:** Calyx of the crinoid *Melocrinites subglobosus* (Weller, 1898); WESM-KCG 19; scale bar 5mm.

Fig. 7 (bottom, opposite). Apatitic skeletal materials fossilized by permineralization. **A:** The conulariid *Conularia milwaukeensis* Cleland, 1911; MPM P28134; scale bar 5mm. **B:** Tooth of the sarcopterygian fish *Onychodus* sp.; MPM 477; scale bar 2mm. **C:** Fin spine of the placoderm fish *Eczematolepis telleri* (Eastman, 1898); MPM 470; scale bar 1cm. **D:** The lingulid brachiopod *Orbiculoidea telleri* Cleland, 1911; MPM P28142; scale bar 5mm. **E:** The lingulid brachiopod *Barroisella? rectilatera* (Hall, 1859); MPM 5136; scale bar 5mm. **F:** Tooth plate from a ptyctodont placoderm fish; UWM-GM 10259; scale bar 5mm.

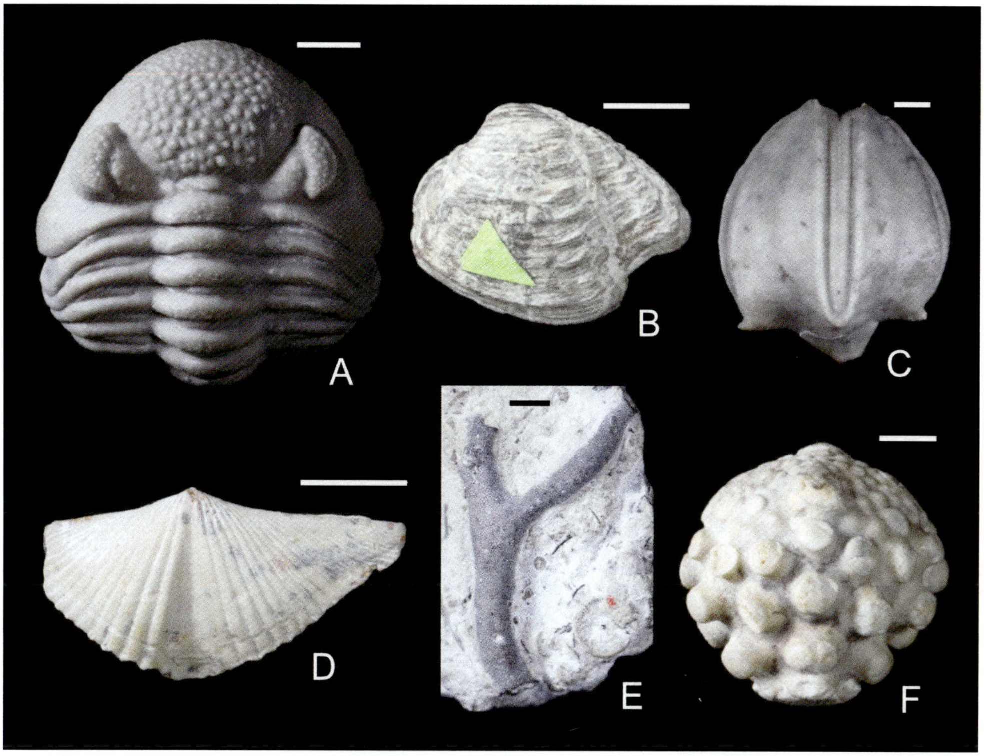
A
B
C
D
E
F

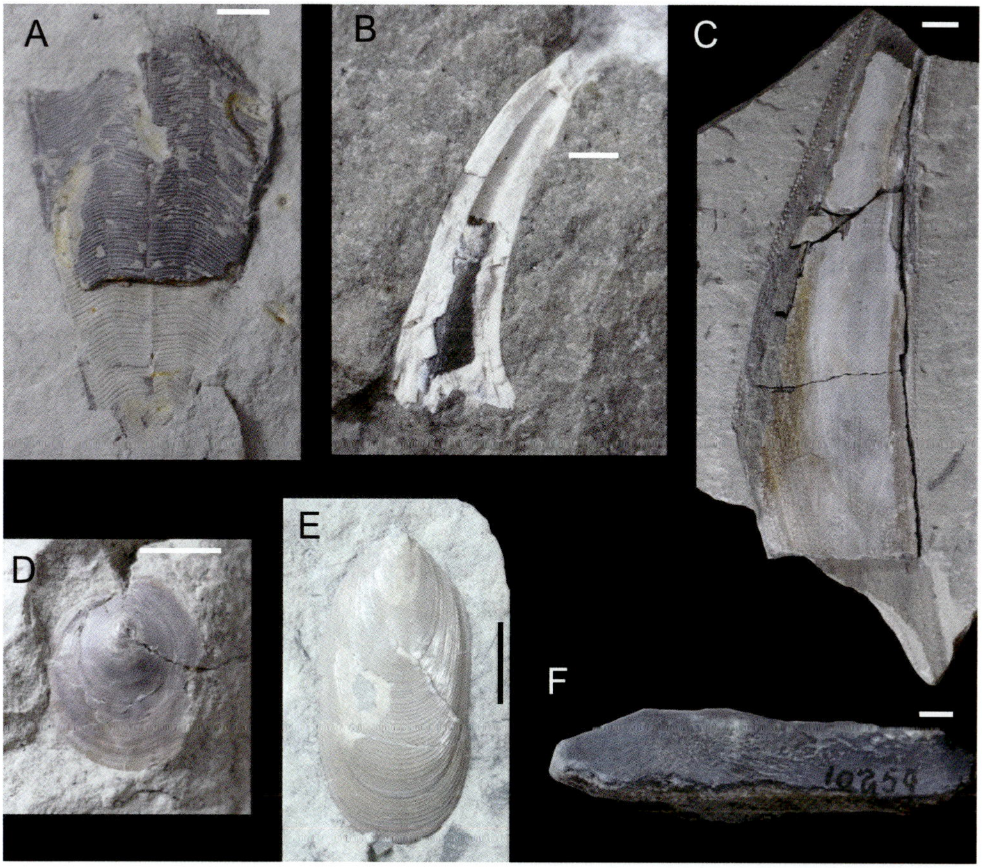
A
B
C
D
E
F

Apatite

The conodonts, conulariids, and linguliform brachiopods of the Milwaukee Formation all contained forms of the phosphatic mineral, apatite. The same is true of the fishes, including the teeth of sarcopterygians and sharks, and the plates, spines, and teeth of placoderms. That apatite persisted after permineralization, imparting an appearance and composition that stand in sharp contrast with that of the surrounding matrix. The colors of these fossils are usually shades of pink, purple, white, black, or amber (Fig. 7). Phosphatic skeletal material of the fish in the Milwaukee Formation is so distinctive in appearance that even small fragments are usually enough to distinguish them from all other fossil taxa.

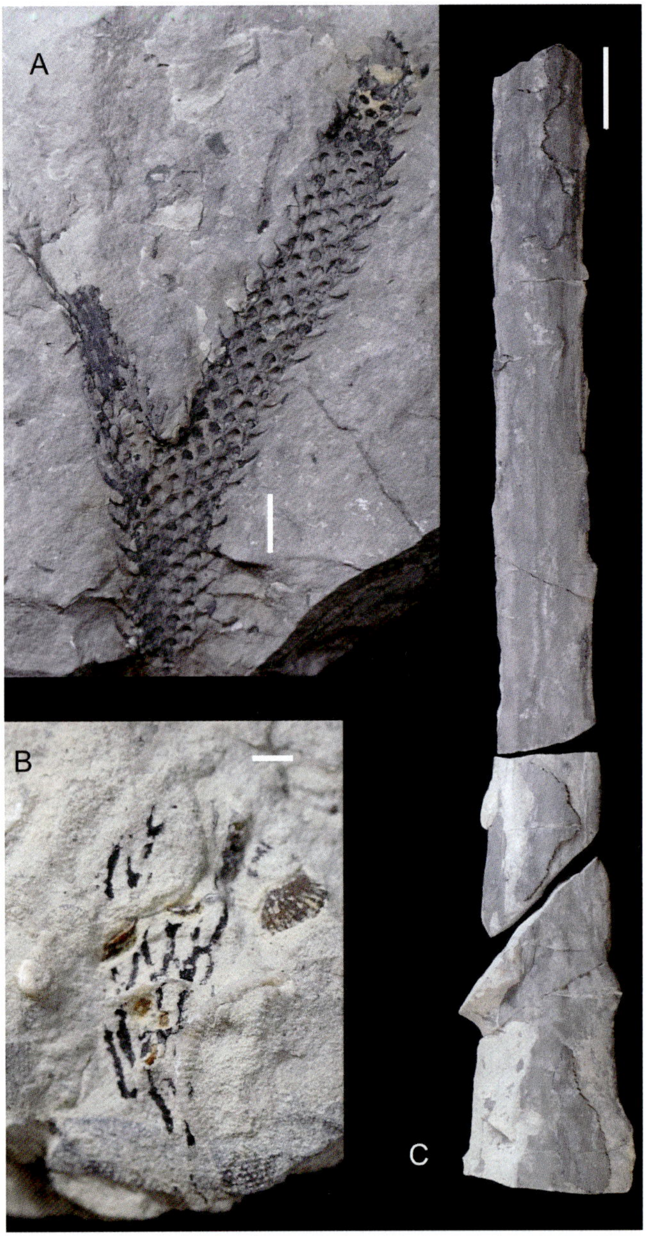

Chitin and Other Organic Materials

The skeletal materials of some organisms lack significant amounts of the previously mentioned minerals, instead consisting mostly of chitin, cellulose or other organic materials. In the Milwaukee Formation, this includes fungi, plant fossils, chitinozoans, phyllocarids, scolecodonts, graptolites, and placoderm scales. As with animals that contain apatite, the color of fossils with organic skeletal remains tends to make them stand out clearly from the surrounding matrix (Fig. 8). Plant fossils, fungi and graptolites tend to be dark gray to black due to their high carbon content. Some of the plant and fungus fossils of the Milwaukee Formation are partially composed of coal (Fig. 9).

Modes of Alteration of Skeletons

Once the organism is buried, the slow process of fossilization can begin. Skeletons are commonly altered in various ways depending upon their original compostions and their tendency to decay, dissolve or have their pores infilled. Common modes of alteration include permineralization, replacement, cabonization, and formation of internal and external molds.

Fig. 8. Organic remains fossilized by permineralization. **A:** Axis (branch) of a lycopod; UWM-GM 9768; scale bar 1cm. **B:** The dendroid graptolite *Callograptus* sp.; MPM P28130; scale bar 2mm. **C:** The arborescent fungus *Prototaxites milwaukeensis* (Penhallow, 1908); MPM 402; scale bar 5cm.

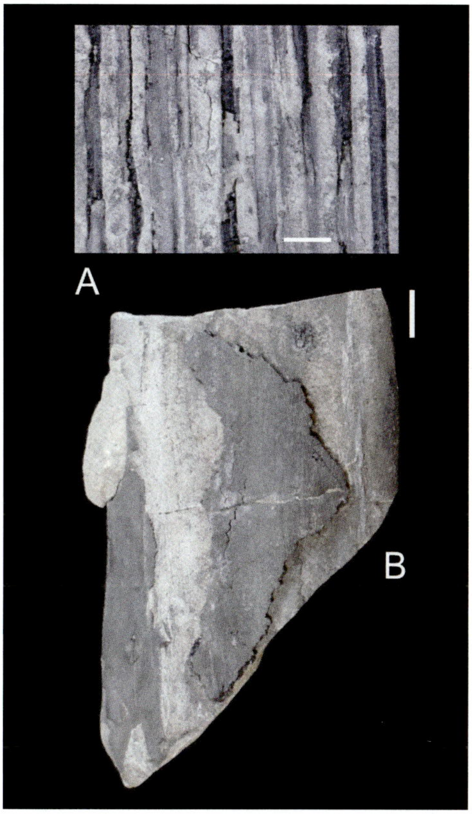

Fig. 9. Fossils of terrestrial organisms partially coalified; Berthelet Member. **A:** Bark from a cladoxylopsid? trunk; UWM-GM 25808; scale bar 1cm. **B:** Piece of the arborescent fungus shown in the previous figure; note the outer layer of coal; scale bar 2cm.

Fig. 10. Valves of the bivalve *Palaeoneilo corrugata* Pohl, 1929, fossilized by replacement by various materials. The valves shown in A, C, and D were presumably replaced by marcasite and certain impurities that then weathered to produce oxidized materials of various colors. The valve shown in B was replaced by silica. **A:** Left valve; MPM 13579; scale bar 5mm. **B:** A silicified left valve; MPM 466; scale bar 2mm. **C:** Right valve; MPM 468; 2mm. **D:** Right valve; MPM 13546; scale bar 2mm.

Permineralization

One fossilization mode found in the Milwaukee Formation, particularly the Lindwurm Member, is permineralization, in which calcite, or to a lesser extent dolomite, silica or other dissolved minerals in the groundwater crystallize within the pores of the remains. This is particularly typical of porous skeletons such as those of echinoderms. The resulting composition of the fossil is therefore a combination of the organism's chemistry (which is quite varied in a biota as diverse as that of the Milwaukee Formation) plus that of the externally derived minerals that crystallized within the organism's remains. This mixture of materials, aided by the effects of post-fossilization chemical changes and weathering processes, helps explain the many variations in colors and physical/chemical properties of the fossils within this formation. The body fossils of most lithified skeletal remains are typically permineraiIized to some degree (Figs. 6–9).

Replacement

Some fossils in the Milwaukee Formation were preserved by replacement, whereby sediment pore water-derived minerals entirely replaced the skeletal materials. This is most conspicuously observed where marcasite, pyrite or silica are the invading minerals. Replacement by marcasite is common in this formation. Many of those fossils are powdery in texture, and of a variety of colors, due to the weathering of marcasite to ferrous sulfate and the introduction of other mineral impurities. In the North Point Member, many of the shells are silicified, with the silica being in the form of white chert (Fig. 10).

Carbonization

This alteration mode, typically occurring in organisms that lack mineralized skeletons, results when the organic materials are reduced mainly to carbon, and is marked by the characteristic black color of the fossil (Figs. 8 & 9).

Internal / External Molds

Many of the fossils in the Milwaukee Formation, especially the Berthelet Member, are in the form of internal and external molds (Fig. 11). These are formed when a shell surrounded by consolidated sediment is dissolved by groundwater or weathered away after the object is buried, thus leaving a gap showing where the shell had been. The layer of rock that surrounds the outside surface of the shell shows the impression of external features known as the external mold. It therefore shows, in reverse, the shell's external dimensions and surface features such as ornamentation. The sediment that filled the concave inside surface of the shell shows, also in inverted relief, the shell's interior dimensions and such structures as muscle scars. The thickness of the shell can be derived by measuring the gap between the internal and external mold. In many cases, the fossil started as a replacement composed of marcasite, but then became molds once the marcasite had altered to ferrous sulfate, which then either dissolved or weathered away. An interesting feature of the internal and external molds of the Berthelet Member is that they generally have a thin, pigmented coating or staining on their surfaces. This generally imparts a charcoal gray or orange coloring, but other colors and shades are common (Fig. 12).

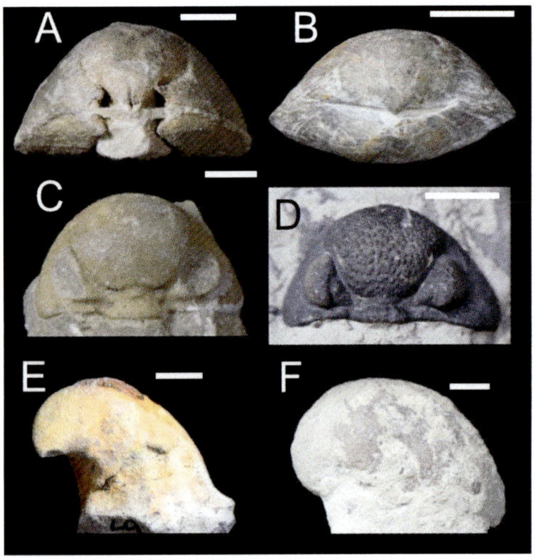

Fig. 11. Internal molds of shells of three types of animals (left), each compared with permineralized shells of the same or similar species (right). Notice that, in contrast to the permineralized equivalent fossils, the internal molds lack tubercles and other surface details, but may include structures indicating the location of features on the interior of the shell. **A:** The brachiopod *Schizophoria iowensis musculosa* Griesemer, 1965; MPM 15232; Berthelet Member; scale bar 1cm. **B:** *Schizophoria iowensis musculosa* Griesemer, 1965; MPM 16435; Lindwurm Member; scale bar 5mm. **C:** The trilobite *Eldredgeops? norwoodensis* (Stumm, 1953); MPM P26724; Berthelet Member; scale bar 5mm; glacial drift; Doctor's Park, Fox Point, Milwaukee County. **D:** *Eldredgeops? norwoodensis* (Stumm, 1953); scale bar 2mm. **E:** The gastropod *Platyceras?* sp.; MPM P28195; scale bar 5mm. **F:** *Platyceras?* sp.; MPM P28131; scale bar 5mm.

Fig. 12. Internal molds of the brachiopod *Eumetabolotoechia greeni* (Cleland, 1911), all recovered from the same small slab. The thin mineral coating on the surfaces of these fossils is more colorful than that of the typical molds from the Berthelet Member. MPM-KCG uncatalogued; scale bar 1cm.

Cavity Formation and Geodization

This occurs in cases where very rapid burial entombed shells of organisms with soft parts intact. After the soft, interior tissues of three-dimensional animal carcasses have decomposed, and if the space they occupied is not filled by sediment, the skeletal remains are essentially hollow. This provides a cavity into which fluids with dissolved minerals, such as calcite, can enter during or after lithification and crystalize on the cavity walls, as occurred often in the Milwaukee Formation. If the invading mineral is calcite, the result is a calcite geode. This was a common process occurring within brachiopods (Fig. 13), bivalves, and crinoid calyces. If massive bryozoans and corals were hollowed by dissolution, they also were targets of geodization. Other dissolved minerals, such as sphalerite (Fig. 14), pyrite (Fig. 15) and millerite (Fig. 16), commonly crystalized within these cavities as well, generally in the center of the void after calcite has crystallized around the margins of the cavities.

In the dolomitic upper layers of the Berthelet Member, these cavities (also known as vugs) may instead be coated or filled with a sticky, black, tar-like substance derived from organic remains that seeped into the cavities from the overlying strata (Fig. 17). The resulting asphaltum vugs are frequently found within partially hollowed cephalopods of this member. Asphaltum also coats the surfaces of some fossil molds (Fig. 18).

Fig. 13. Two views of a geodized brachiopod filled with calcite; MPM P28132; Lindwurm Member; collected by KCG; scale bar 5mm.

Fig. 14. Geodized brachiopod with a single octahedral crystal of sphalerite (zinc sulfate) and calcite lining; MPM P28133; Lindwurm Member; collected by KCG; scale bar 5mm.

Fig. 15. Crystals of pyrite and calcite in a partially cavitated cephalopod; UWM-GM 32967; Berthelet Member; scale bar 1cm.

Fig. 16. Millerite (nickel sulfide) secondarily altered on its surface to emerald green zaratite (hydrous nickel carbonate) and calcite in a cavitated bryozoan; MPM Mi 20590; probably Berthelet Member; collected by KCG; scale bar 5mm.

Fig. 17. Asphaltum in an unidentified cavity; MPM Mi 20591; Berthelet Member; collected by KCG; scale bar 5mm.

Fig. 18. Internal mold of the brachiopod *Orthospirifer milwaukeensis* (Cleland, 1911) covered with asphaltum; MPM 10978; Berthelet Member; collected by IE; scale bar 1cm.

Preservation of Color Patterns

Certain species from the Milwaukee Formation often preserve color patterns (Fig. 19). This mostly occurs with cranaenid and lingulid brachiopods, although it has also been observed with phacopid and proetid trilobites and the fin spines of certain fish, such as *Gamphacanthus*. These are typically seen as bands or spots with contrasting shades or hues. While there is little dispute that such patterns seen on brachiopod fossils are original surface features, there is some controversy regarding trilobites. It has been suggested that the color patterns of some trilobites may not have been visible from the surface of the living animal, and may instead show the placement of other structures, such as muscle attachments, slightly below the surface of the exoskeleton (Babcock, 1982). Other researchers interpret the spots and color bands of certain Devonian trilobites to be the original patterns that may have served as camouflage (McRoberts et al., 2013; Schoenemann et al., 2014). Due to the many chemical changes that take place during and after fossilization, in no cases are the actual colors of these patterns likely to be the same as when the animals were alive.

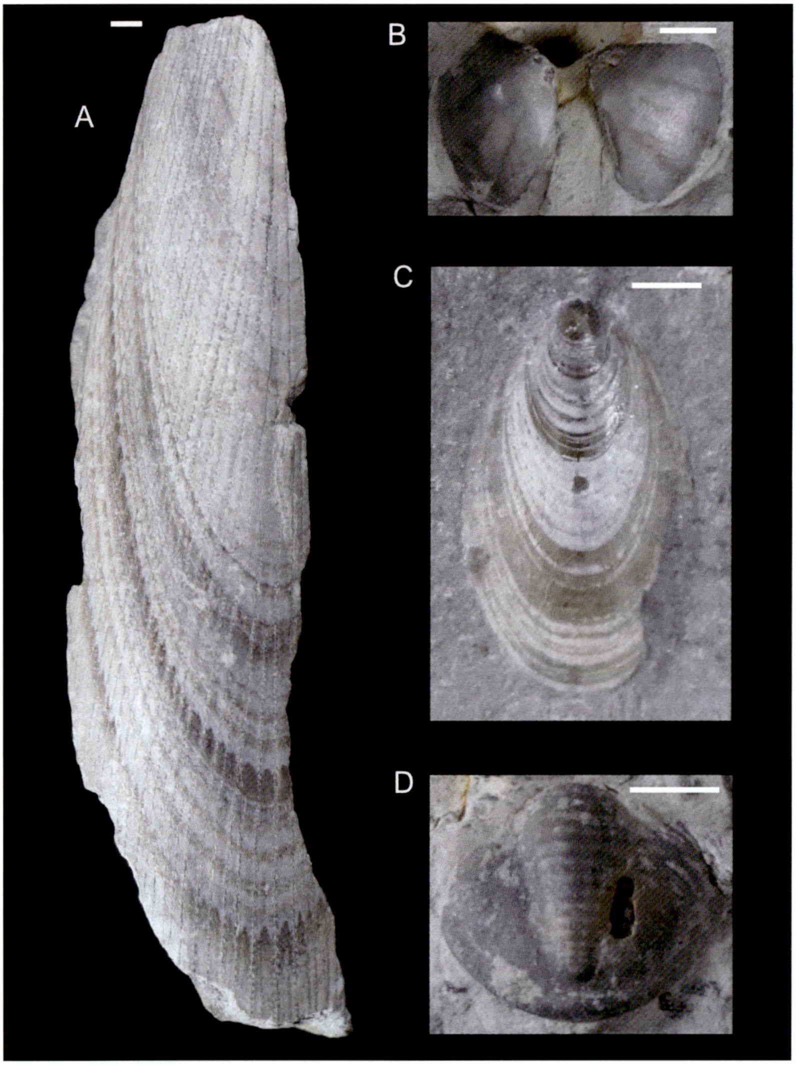

Fig. 19. Color patterns in skeletal materials. **A:** Fin spine from the ptyctodont *Gamphacanthus politus* (Newberry, 1889); UWM-GM 18269; Berthelet Member; scale bar 1cm. **B:** Splayed valves of the brachiopod *Cranaena* sp.; WESM-KCG 1115; Lindwurm Member; collected by LVS; scale bar 5mm. **C:** Brachial valve of the brachiopod *Barroisella? milwaukeensis* (Cleland, 1911); MPM 236; North Point Member; scale bar 1mm. **D:** Pygidium from a proetid trilobite; WESM-KCG 1049; Lindwurm Member; collected by JSG; scale bar 5mm.

Trace Fossils

Trace fossils are evidence of the movements or other behavior of an organism, as opposed to the organism itself. Most of the trace fossils of the Milwaukee Formation are in the form of burrows in the sediment. *Palaeophycus*, *Chondrites*, and *Zoophycos* are among its more common ichnogenera. These are discussed in more detail in Chapter 23.

Life and Death Assemblages

Throughout the Milwaukee Formation are layers that are packed with fossils on their surfaces (Figs. 20 & 21). These fossils are so prolific and well preserved that many observers enjoy imagining that they are viewing a snapshot of a seafloor with its natural assemblage of living creatures. Except for the color discrepancies we discussed above and the fact that all soft-bodied creatures have been removed from the

Fig. 20. Associated pieces of brachiopods, bryozoans, crinoid columnals, and trilobites. Such associations are typical of certain layers in the Lindwurm Member. MPM P26725; collected by LVS; scale bar 1cm.

Fig. 21. Slab containing numerous fragments of the calcareous skeletal materials from brachiopods, bryozoans, hederelloids. crinozoans, corals, and trilobites. WESM-KCG 1111; Lindwurm Member; collected by KMG; scale bar 5mm.

scene, many of these examples of fossilized “mini-ecosystems” may actually appear fairly similar to the way they looked in life. However, we must remember that weeks, months or years may have transpired before they were buried. During that time currents made them subject to minor transportation, redistribution, sorting and reorientation. In addition, remains of different generations and in some cases slightly different communities could accumulate in a single bed of fossils; this process has been termed “time-averaging”.

Various remains in these assemblages may have been transported—the question being to what degree. We thus look for clues. For example, we know that *Aulopora, Petrocrania* and *Microconchus* that are still cemented to other shells or massive bryozoan colonies were obviously not transported away from their hosts. In addition, complete remains of delicate skeletons such as articulated crinoids and trilobites have almost certainly been buried rapidly with little to no transport. Other lines of evidence are not as conclusive. Shells that are severely broken and worn were probably transported more than ones that were fossilized intact; however, the gases within a nautiloid shell can enable it to float for miles without breakage or appreciable wear. The exact location where that cephalopod lived and died is, therefore, difficult to ascertain. The plants and fungi fossilized in the Milwaukee Formation certainly lived on land. The more delicate, well preserved ones probably did not live too far from where they were preserved, while others might have floated in with the currents from many miles away. As with cephalopod shells and planktonic organisms, it may not be possible to determine their exact place of origin or the distance they travelled.

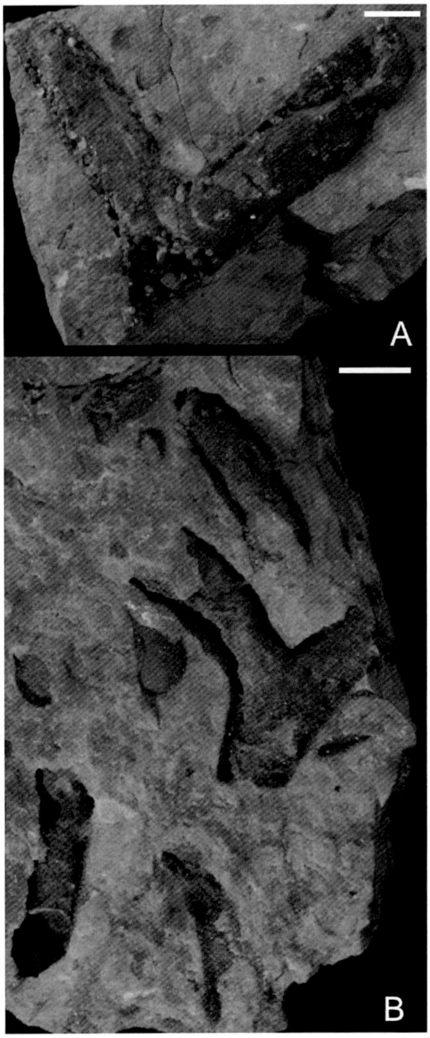

Quality of Preservation

While some fossils of the Milwaukee Formation can be very well preserved, showing fine details, non-mineralized tissues, or color patterns, others can be so poorly preserved as to be unidentifiable. This is especially true of the bryozoans and corals. Most of the interiors of those fossils from the Milwaukee Formation have been obliterated by dissolution and/or dolomitization (Fig. 22), thus obscuring important diagnostic features and causing identification to rely only on external features (Cleland, 1911).

As with any other problem being studied in paleontology and other sciences, what we learn is only as good as the totality of the evidence accumulated. For that reason, we must consider all lines of evidence and study as many fossils as we can in order to best describe and explain these ancient organisms.

Fig. 22. Apparent ramose bryozoan zoaria; Berthelet Member; scale bars 1cm. Dissolution and recrystallization have obscured all internal and external identifying features. **A:** UWM-GM 9577; **B:** UWM-GM 9576.

Chapter 5 — Foraminifers, Ostracods, and Other Microorganisms

Many major taxa of microorganisms are sure to have lived in Devonian Milwaukee, while most were either not preserved or their fossils have been overlooked. Among the occurrences from the Milwaukee Formation for which there is a record, but a poor one, are the unicellular, siliceous **radiolarians** and flask-shaped, enigmatic **chitinozoans**. Detectability, size, ease of preparation or extraction, and state of preservation of fossils largely dictate which taxa receive the most attention, and this may be as great a factor in their scanty documentation as is their rarity or absence. Better studied microfossils from the Milwaukee Formation are the agglutinated foraminifers and ostracods.

Foraminifers (Protista: Foraminiferida)

One of the best-studied groups of microorganisms found in the Milwaukee Formation is the unicellular Foraminifera, even though only one comprehensive study has been conducted and it was never published (Roberts, 1972). All of these foraminifers are of the agglutinated type, having skeletons produced by adhering various organic or inorganic particles to the organism's organic test, as opposed to secreting the entire skeleton. Roberts (1972) recognized about ten species of agglutinated foraminifers from the Lindwurm and/or Berthelet Members (Fig. 23)—all presumed to have lived a sessile, benthic mode of life, as they are either found attached to other organisms or show impressions suggesting previous attachment. Mayer (2001) also reported agglutinated foraminifers from the North Point Member.

Ostracods (Arthropoda: Crustacea: Ostracoda)

The kidney bean-shaped, bi-valved, mostly free-swimming crustaceans known as ostracods are the earliest-documented and one of the most abundant fossils of microorganisms in the Milwaukee Formation. However, because ostracods are calcareous, they are difficult to clean and segregate from the calcareous matrix. Specimens suitably prepared for producing acceptable photographs were therefore not available for inclusion in this book. Cleland described three taxa from this formation in his 1911 book; Raasch (1935) listed considerably more, as follows: *Amphissites* spp., *Bollia ungula*, *Cytherella*, *Dizygopleura*, *Glyptopleura*, *Halliella*, *Healdia* spp., *Hollina* sp., *Jonesina* spp., *Kirkbya*, *Octonaria*, *Paraechmina*, *Treposella lyoni*, and *Ulrichia conradi*.

Fig. 23 (next page). Agglutinated foraminifers; collected by JER; scale bar 0.5mm; alphameric string after "UWM-GM" is the sample number. Images are photographs of selected images from Roberts, 1972. **A:** *Fairiella carmani* Summerson, 1958; UWM-GM 3b/11B; figured in Roberts, 1972, Pl 2, Fig 1. **B:** *Webbinelloidea similis* Stewart & Lampe, 1947; UWM-GM 3b/C-1B; figured in Roberts, 1972, Pl 1, Fig 17. **C:** *Psammosphaera elongata* Summerson, 1958; UWM-GM 3a/A1; figured in Roberts, 1972, Pl 1, Fig 6. **D:** *Tolypammina* sp. C; 3b/C-7; figured in Roberts, 1972. Pl 2, Fig 7. **E:** *Oxinoxis ligula* Gutschick, Weiner & Young, 1961; UWM-GM 3d/WFB-1; figured in Roberts, 1972, Pl 3, Fig 7. **F:** *Tolypammina* sp. A; UWM-GM 3c/A-1; figured in Roberts, Pl 2, Fig 3. **G:** *Rhabdammina* sp.; UWM-GM 3c/A-1; figured in Roberts, 1972, Pl 1, Fig 1. **H:** *Psammosphaera delicatula* Stewart & Lampe, 1947; figured in Roberts, 1972, Pl 1, Fig 3. **I:** *Reophax* sp.; UWM-GM 3b/C-11A; figured in Roberts, 1972, Pl 2, Fig 13. **J:** *Tolypammina* sp. B; UWM-GM 3b/C-2; figured in Roberts, 1972, Pl 2, Fig 4.

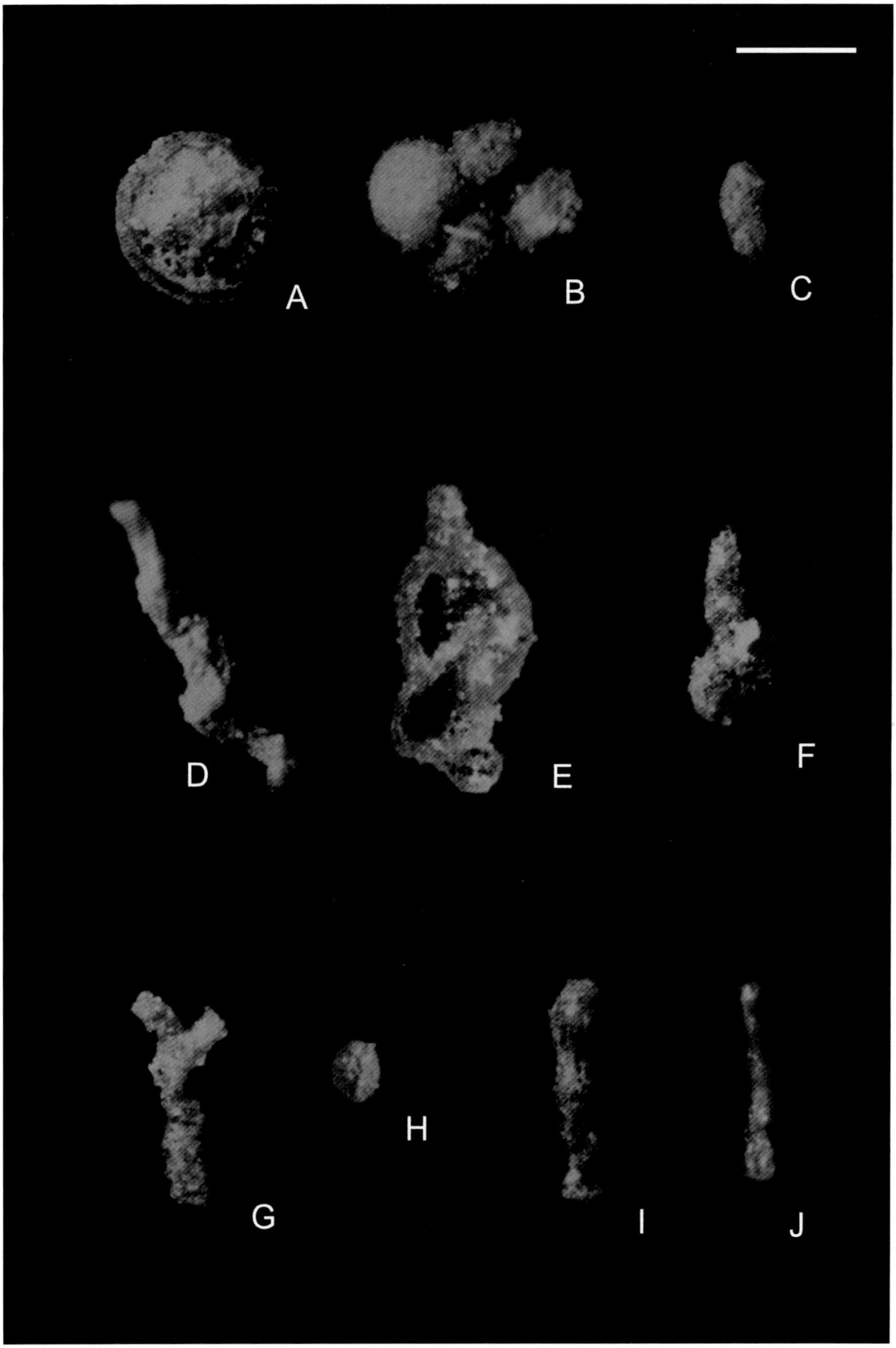
A
B
C
D
E
F
G
H
I
J

Chapter 6 — Conulariids
(Cnidaria: Scyphozoa: Conulariida)

The Conulariida (Ediacaran–Triassic) was a group of sessile benthic, polypoid scyphozoan cnidarians (Ford et al., 2016; Van Iten et al., 2006b; Van Iten et al., 2014; Van Iten et al., 2016a; Van Iten & Südkamp, 2010) having a steeply pyramidal, generally four-sided periderm constructed of alternating, extremely thin organic and apatitic lamellae that in some species could be folded to cover the oral end of the animal. In nearly all species, the corners of the periderm are sulcate, and the faces are crossed by straight or (more commonly) adorally arching transverse ridges (ribs), often bearing small nodes, or by transverse rows of nodes (Fig. 24). The midline of each face may be marked by a sulcus or by offset of some or all of the transverse ribs, and in many conulariids it is the site of a continuous or seriated, internal carina or pair of carinae. An internal carina, continuous or seriated, may also be present at the corners. The apical end of the periderm is normally tapered to a near point, but some specimens terminate instead in a more or less smooth, outwardly convex apical wall (schott) that may have formed in response to mechanical severance of the body.

Because of their thin, originally flexible periderm, conulariid fossils commonly are flattened to a high degree, and this is especially true of *C. milwaukeensis*, most specimens of which are essentially collapsed along the two corners lacking an internal carina, thus showing only two of their four faces (Fig. 25).

Fig. 24. The conulariid *Conularia milwaukeensis* Cleland, 1911, with selected morphological features labeled; MPM P28134; Berthelet Member. **A:** Two faces broken at oral and apical extremities; scale bar 5mm. **B:** Close-up view of periderm. **C:** Close-up of impression of periderm with nodes shown here as small, linearly arranged pits along the ribs.

Conularia milwaukeensis Cleland, 1911, found in the Lindwurm and Berthelet Members, is the only species of conulariid identified with certainty from the Milwaukee Formation. It is also found in the Little Cedar Formation of Iowa. Van Iten et al. (2006a) gave a detailed, modern description of this species. These authors discovered that *C. milwaukeensis* exhibits a continuous, internal carina at just two of its corners. The carinate corners are situated on opposite sides of the plane connecting the two non-carinate corners, and in flattened specimens it is the non-carinate corners that have been brought together. In addition to the two internal carinae, some specimens also exhibit a schott.

C. milwaukeensis has also been included in other recent studies (Leme, 2006; Babcock & Feldman, 1986). Cleland (1911) listed additional species from the Milwaukee Formation that he assigned to the conulariids, but those have since been reassigned to other invertebrate groups, and are discussed in the appropriate chapters of this book. The genus *Sphenothallus*, which may have been closely related to conulariids, has also been reported from the Milwaukee Formation (Van Iten et al., 2006a).

Fig. 25. *Conularia milwaukeensis* Cleland, 1911, Conulariida; Berthelet Member; scale bar 5mm. **A:** MPM 245; figured in Cleland, 1911, Pl 26, Fig 4; collected by CEM. **B:** MPM P28134; referenced in Van Iten et al., 2006 under the number 1045; collected by KCG. **C:** MPM 244; figured in Cleland, 1911, Pl 24, Fig 6; collected by CEM. **D:** MPM 20252; figured in Babcock & Feldman, 1986, Figs 7.7 & 7.8; collected by ED. **E:** PAL 85988; holotype; figured in Cleland, 1911, Pl 26, Fig 4; Courtesy of Smithsonian Institution, Department of Paleobiology, National Museum of Natural History. **F:** UWM-GM 5390.

Chapter 7 — Rugose Corals
(Cnidaria: Anthozoa: Rugosa)

Rugose corals (Ordovician–Permian) were sessile, marine, calcareous, solitary and colonial anthozoan cnidarians distinguished from each other largely by features relating to their septa. Their exoskeletons were usually cone-shaped, narrowing downward toward their apex, where they attached to shells or other objects (Fig. 26). When alive, they are likely to have had an anemone-like appearance, including soft mesenteries (located between the calcareous septa) that partition the body, and a central mouth situated at the top surrounded by tentacles (Hill, 1981a).

The solitary species, *Odontophyllum* cf. *O. convergens* (Hall, 1882) (Figs. 27 & 28) and *Heliophyllum* cf. *H. corniculum* (Lesueur, 1820), are the only rugose corals recognized thus far from the Milwaukee Formation. Many specimens attributed to *O. convergens* were collected after Cleland wrote his book and now reside at the Milwaukee Public Museum; however, most are in the form of internal molds and therefore do not permit study of the interior, thus making definitive identification impossible. Rugose corals are relatively rare in the Milwaukee Formation.

Most of the rugose coral specimens from the Berthelet Member in museum collections are only fillings of the calice. The calice was the distal surface of the corallite, generally bowl-shaped. Consequently, those fossils do not have the form of the corallite, but of the base of the polyp.

Fig. 26. Lateral view of filling of calice of *Odontophyllum* cf. *O. convergens* (Hall, 1882), Stauriida: Hallidae; UWM-GM 1716; Berthelet Member; scale bar 5mm.

Fig. 27. Ventral and lateral views of filling of calice of *Odontophyllum* cf. *O. convergens* (Hall, 1882), Stauriida: Halliidae; UWM-GM 1719; Berthelet Member; scale bar 5mm.

Fig. 28. Ventral and lateral views of filling of calice of *Odontophyllum* cf. *O. convergens* (Hall, 1882), Stauriida: Halliidae; UWM-GM 1697; Berthelet Member; scale bar 5mm.

Chapter 8 — Tabulate Corals
(Cnidaria: Anthozoa: Tabulata)

Tabulate corals were almost exclusively Paleozoic, marine, calcareous, colonial, anthozoan cnidarians characterized by their exoskeleton of slender tubes (corallites) crossed by many transverse partitions known as tabulae (Hill, 1981b) (Fig. 29). Some tabulates were important reef-builders in the Silurian and Devonian Periods; however, there is no evidence of reef development in the Milwaukee Formation.

As indicated earlier, most of the fossil corals of the Milwaukee Formation lack details of their interior, thus making identification difficult. Nonetheless, remaining external features are sufficient to identify several reptant, encrusting, and ramose species. In addition to the species shown here (Figs. 30–32), Cleland (1911) listed and included drawings of the following taxa from the Milwaukee Formation: *Aulopora serpens* Goldfuss, 1829, *Aulocystis agglomerata* (Grabau, 1901), *A. flabellata* (Greene, 1902), *A. intermedia* (Nicholson, 1899), *A. jacksoni*? (Grabau, 1899), *Favosites alpenensis*? Winchell, 1866, *F. radiatus*? Rominger, 1876, and *Striatopora iowensis* (Owen, 1840).

Most of these appear to be present in the museum collections studied, but are not sufficiently prepared for photographic purposes. It should be noted that the species assigned here to *Aulocystis* were assigned by Cleland to *Ceratopora*, but the species representing that genus had already been given the valid name *Aulocystis* in 1885. Consequently, the name *Ceratopora* is invalid. Larger colonial corals that are common in the Middle Devonian rocks of many other Midwestern localities, such as the Coralville Formation of Iowa, are essentially absent from the Milwaukee Formation.

Fig. 29. The reptant tabulate coral *Aulopora* cf. *A. annectens* Clarke, 1885 with selected external morphological features labeled; MPM P28744 A; scale bar 1cm.

Fig. 30. *Aulocystis* sp., Auloporida: Aulocystidae; MPM P28136; Berthelet Member; collected by KCG; scale bar 5mm.

Fig. 31. Cephalopod encrusted by *Planalveolitella monroei* (Cleland, 1911), Favositida: Alveolitidae; PAL 78315; holotype; Berthelet Member; figured in Cleland, 1911, Pl 2, Figs 1&2; collected by EET; scale bar 1cm. Courtesy of Smithsonian Institution, Department of Paleobiology, National Museum of Natural History.

Fig. 32. *Aulocystis rugosa* (Cleland, 1911), Auloporida. Aulocystidae; PAL 99495; holotype; Lindwurm Member; figured in Cleland, 1911, Pl 21, Fig 6; collected by EET; scale bar 5mm. Courtesy of Smithsonian Institution, Department of Paleobiology, National Museum of Natural History.

Chapter 9 — Tentaculitoids
(Lophophorata?: Tentaculitoidea)

Consisting of the suborders, Tentaculitida, Microconchida, and Cornulitida, the tentaculitoids were enigmatic, generally marine animals that bore a conical or spiral, calcareous shell. With little known of their soft parts and lacking a molecular biological profile, their zoological affinities are uncertain. Based mainly on shell morphology, ontogeny, and the meager record of their soft parts, these animals have been assigned at various times to numerous groups, including mollusks, annelids, and in the case of the cornulitids, to cnidarians. Some of the more recent studies suggested an alignment with lophophorates—the lineage that includes the phoronids, bryozoans, and brachiopods, grouping the tentaculitids, microconchids and cornulitids together as tentaculitoids (Vinn & Mutvei, 2009; Vinn & Zaton, 2012).

Tentaculitida (Cambrian–Devonian)

This was the order of the Tentaculitoidea having shells that were typically small (1–80mm), conical in shape, and had transverse rings and, in some species, transverse and/or longitudinal striations—the transverse ones being growth lines (Fig. 33). The shell walls were usually laminated in two to five layers; thick walled forms were pierced by tiny radial canals (Fisher, 1962). Showing no evidence of burrowing into sediment or attaching to other objects, but generally having a thick, ringed shell, a vagile benthic lifestyle is more likely than an infaunal, attached or nectonic one. *Tentaculites bellulus* is commonly found throughout the Milwaukee Formation, often in clusters (Figs. 34 & 35). This clustering may have been enhanced by post-mortem sorting by shell size, shape and density.

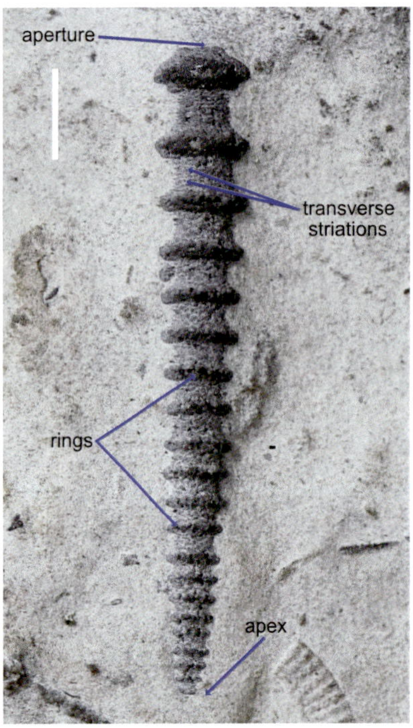

Fig. 33. Large specimen of the tentaculitid *Tentaculites bellulus* Hall, 1879 with selected external morphological features labeled; MPM 248; scale bar 2mm.

Fig. 34. *Tentaculites bellulus* Hall, 1879, Tentaculitida: Tentaculitidae; Lindwurm Member. **A:** MPM 22973; collected by BB; scale bar 5mm. **B:** MPM 248; collected by CEM; scale bar 2mm.

Fig. 35. *Tentaculites bellulus* Hall, 1879, Tentaculitida: Tentaculitidae; MPM 247; Lindwurm Member; figured in Cleland, 1911, Pl 4, Fig 6; collected by CEM; scale bar 5mm.

Microconchida (Ordovician–Jurassic)

The microconchids were encrusting epibionts with a small spiral shell. Until observations on the shell microstructure led to the present placement (Zaton & Vinn, 2011), species now assigned to Microconchida were thought to be polychaete annelid worms, specifically in the family Spirorbidae. One such species found in the Milwaukee Formation and certain other Devonian formations was for decades named *Spirorbis omphalodes* (Goldfuss, 1831), but Taylor & Vinn (2006) have since reassigned this species to *Microconchus* (Fig. 36). Another likely microconchid from the Milwaukee Formation is *Anticalyptraea apicatus* (Cleland, 1911) (O. Vinn, personal communication, 13-Nov-2017).

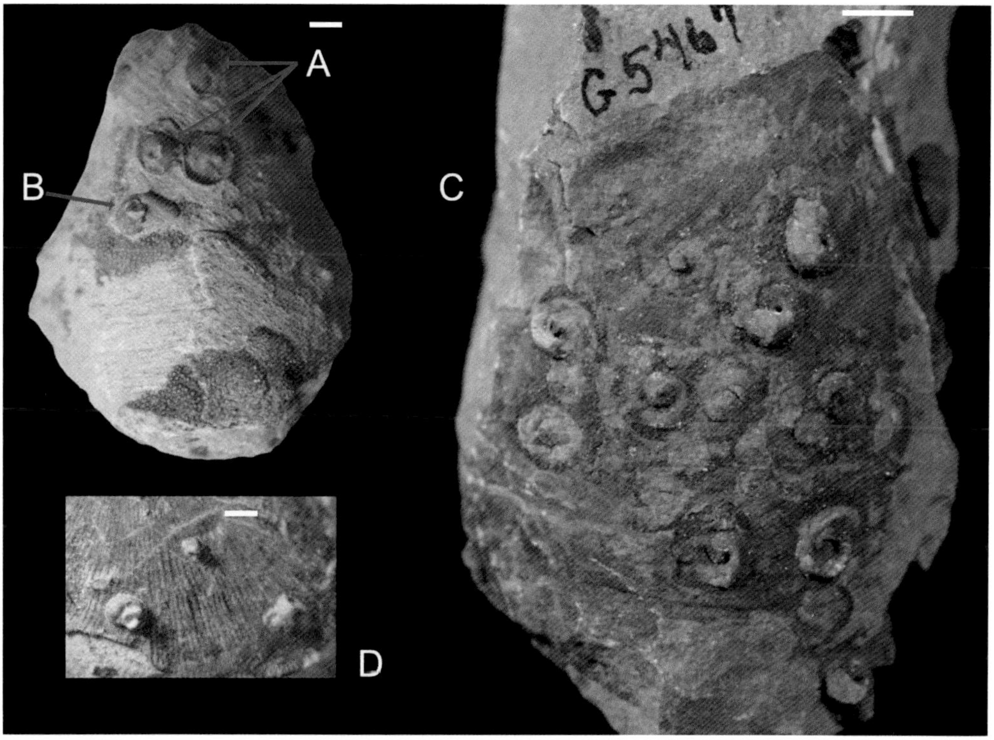

Fig. 36. Two species of microconchids. **A:** Three specimens of *Anticalyptraea apicatus* (Cleland, 1911), Microconchida: Microconchidae, attached to a bryozoan zoarium; MPM P28193; Lindwurm Member; collected by GOR; scale bar 2mm. **B:** *Microconchus* sp., Microconchida: Microconchidae; MPM 15487; Lindwurm Member; collected by GOR; scale bar 2mm. **C:** This group of unusually large individuals of *Microconchus* sp. is attached to a cephalopod; UWM-GM G5467; Berthelet Member; scale bar 5mm. **D:** Three specimens of *Microconchus* sp. attached to a brachiopod; MPM P28179; Lindwurm Member; scale bar 2mm.

Cornulitida (Ordovician–Pennsylvanian)

While the larvae of this order were probably free-swimming, adult cornulitids were encrusters, like the microconchids. As the larvae developed, they attached to other objects where the animals remained for the rest of their lives. Ring development was strongest during the intermediate growth stages, after which the rings became obscure. Cleland (1911, Pl. 4, Fig. 2) attributed a specimen from the Lindwurm Member to *Cornulites* sp. Although this specimen cannot be located, and no other unequivocal specimens have been found or cataloged in the MPM or Greene collections, other examples of this small form may turn up with diligent examination of those and other collections that hold material from the Milwaukee Formation.

Chapter 10 — Bryozoans
(Lophophorata: Bryozoa)

The Bryozoa (Ordovician–Holocene) is the phylum of lophophorate animals comprised of exclusively colonial, aquatic, generally sessile metazoans, regarded as coelomate, with a retractable food-gathering lophophore, U-shaped digestive tract and typically calcareous skeleton (Boardman et al., 1983). From a macroscopic perspective, bryozoan colonies (i.e., zoaria) have a wide range of growth habits, such as massive, encrusting, branching, fenestrate, and reptant, sometimes with more than one of these forms present within the same colony. Their surfaces are covered by minute, closely spaced apertures defined by the outline of the zooidal walls (Fig. 37). The zooids of bryozoan taxa can be polymorphic (Fig. 38), whereby each polymorph is specialized to provide a specific function, such as reproduction, support, cleaning or defense. The zoarium tends to be more complex than coral exoskeletons; it includes the calcareous layers of all of the polymorphs in addition to those of the normal feeding zooids and other calcareous structures within the zoarium. Many of these structures can only be observed under magnification, often requiring thin sections for examination of the interior structures. The placement of bryozoans in either the Deuterostomia or Protostomia is ambiguous based on embryonic and molecular considerations.

The bryozoans of the Milwaukee Formation, like the corals, typically lack preservation of interior structures and, in the case of specimens from the Berthelet Member, details of external anatomy. Distinguishing fossilized bryozoan taxa from each other, and even from unrelated colonial organisms such as corals, can therefore be difficult. Several of the many bryozoan species from that formation are illustrated here (Figs. 39–52). Bassler (*in* Cleland, 1911) included the following species as well: *Condranema monroei* (Bassler *in* Cleland, 1911), *Cyclotrypa collina* (Ulrich, 1890), *Fistulipora monticulata* Ulrich, 1890, *F. romingeri* Nicholson & Foord, 1885, *F. utriculus* Rominger, 1866, *Hemitrypa tenera* Ulrich, 1890, *Petalotrypa compressa* Ulrich, 1890, *Reteporina hamiltonensis* (Prout, 1866), *Rhopalonaria tenuis* Ulrich & Bassler, 1904, and *Semicoscinium rhombicum* Ulrich, 1890.

Cleland also listed *Allonema fusiforme* (Nicholson & Etheridge, 1877), *Ascodictyon floreale* Ulrich & Bassler, 1904, and *A. stellatum* Nicholson & Etheridge, 1877; however, Wilson & Taylor (2014) and Jarachowski et al. (2016) provided evidence suggesting that those genera are not bryozoans and are best considered *incertae sedis* (Microproblematica). Cleland also included *Vinella*? *devonica* (Bassler *in* Cleland, 1911) as a Milwaukee Formation bryozoan, but Mark Wilson stated that *Vinella* also appears to lie outside of the Bryozoa (personal communication, 18-Aug-2018).

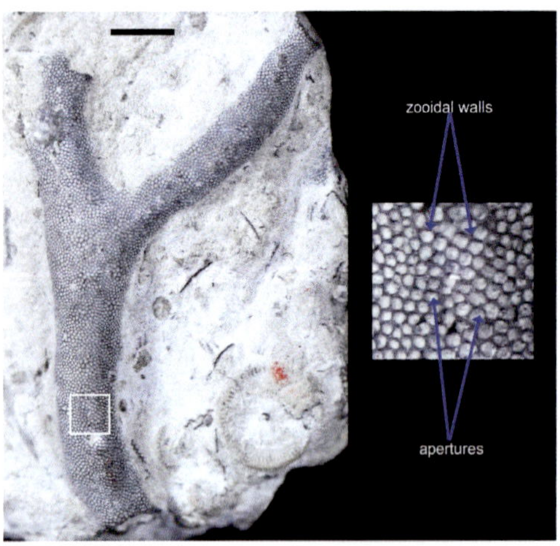

Fig. 37. A zoarium of the ramose bryozoan *Eridotrypa appressa* (Ulrich, 1890) with selected external morphological features labeled; UWM-GM 5880; scale bar 5mm.

Fig. 38. Branching zoarium of *Sulcoretepora hamiltonensis* (Ulrich, 1890). Note the subspherical structures which may be brooding chambers. KCG collection; Lindwurm Member; collected by DKG; scale bar 2mm.

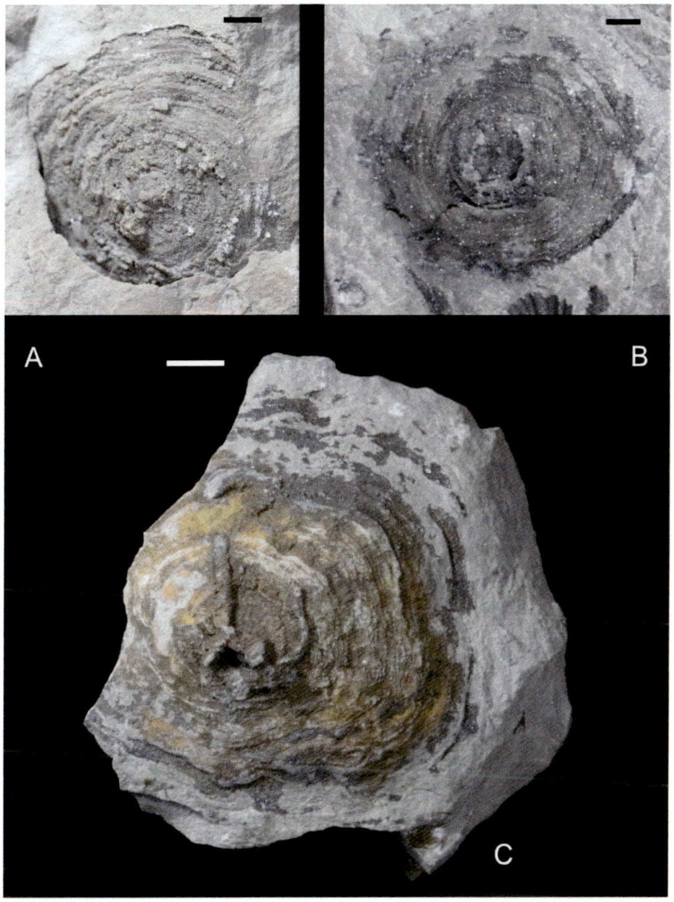

Fig. 39. *Lichenalia concentrica* Hall, 1851, Cystoporata: Rhinoporidae; Berthelet Member. **A:** MPM P28180; collected by KCG; scale bar 5mm. **B:** UWM-GM 9644; Whitefish Bay?; scale bar 5mm. **C:** MPM P28181; collected by KCG; scale bar 1cm. These represent the lower surface of the zoarium, showing the concentric growth lines.

Fig. 40. *Sulcoretepora hamiltonensis* (Ulrich, 1890), Cystoporata: Cystodictyonidae; MPM 16181; Lindwurm Member; collected by GOR; scale 1cm.

Fig. 41. *Sulcoretepora hamiltonensis* (Ulrich, 1890), Cystoporata: Cystodictyonidae; UWM-GM 2006; Lindwurm Member; scale bar 2mm.

Fig. 42. *Sulcoretepora hamiltonensis* (Ulrich, 1890), Cystoporata: Cystodictyonidae; MPM P28137; Lindwurm Member; collected by KCG; scale bar 1cm.

Fig. 43. Two views of a large zoarium of *Leioclema minutissimum* (Nicholson, 1875), Trepostomatida: Leioclemidae; MPM P28138; Lindwurm Member; collected by KCG; scale bar 2cm.

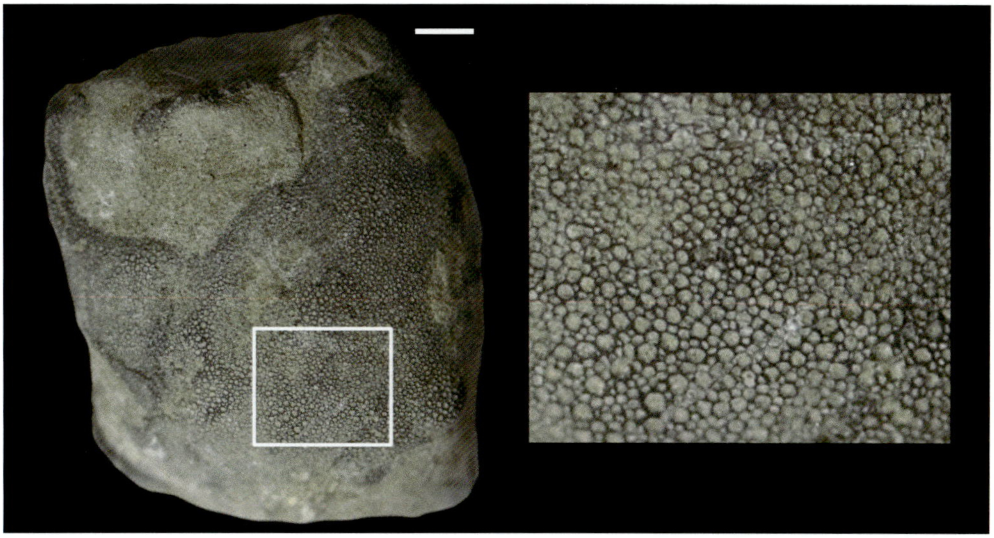

Fig. 44. *Leioclema minutissimum* (Nicholson, 1875), Trepostomatida: Leioclemidae; MPM 16147; Lindwurm Member; collected by GOR; scale bar 2mm for the zoarium.

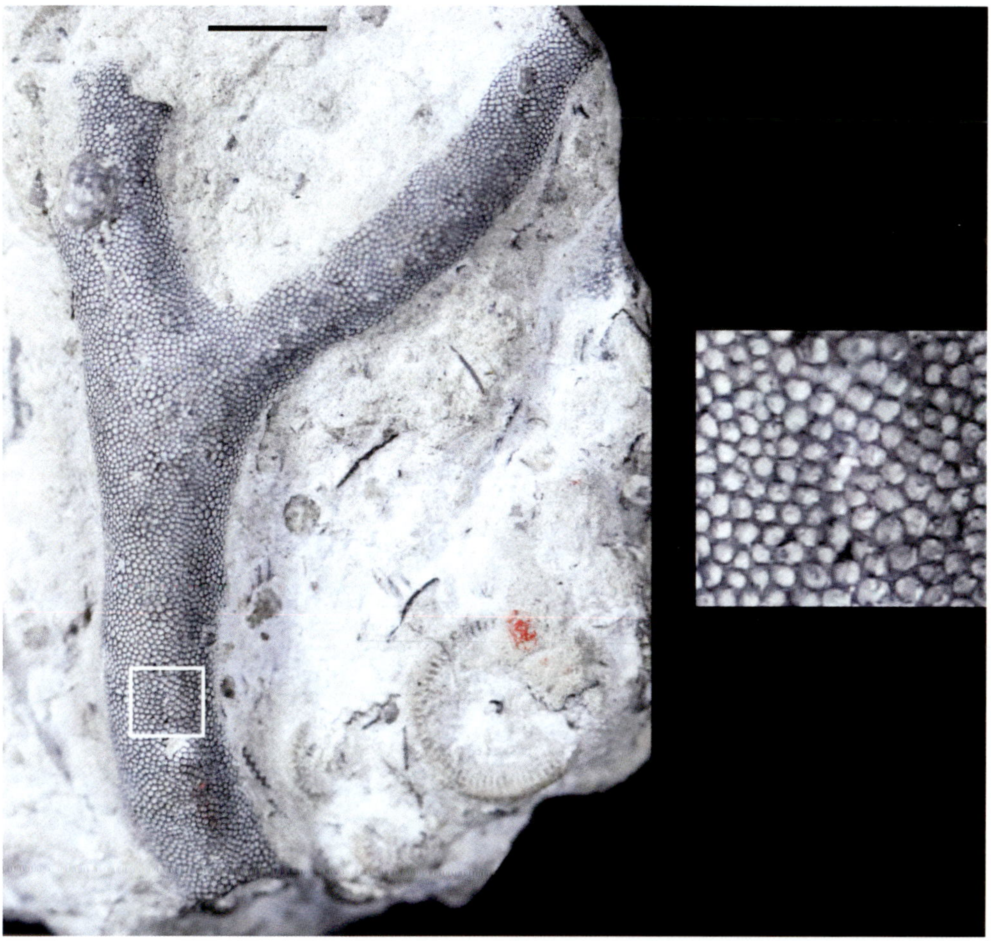

Fig. 45. *Eridotrypa appressa* (Ulrich, 1890), Trepostomatida: Aisenvergiidae; UWM-GM 5880; Lindwurm Member; scale bar 5mm for the zoarium.

Fig. 46. *Eridotrypa appressa* (Ulrich, 1890), Trepostomatida: Aisenvergiidae; MPM 15874; Lindwurm Member; collected by GOR; scale bar 2mm for the zoarium.

Fig. 47. *Fistulipora* cf. *F. romingeri* Nicholson & Foord, 1885, Cystoporata: Fistuliporidae; MPM 15565; Lindwurm Member; collected by GOR; scale bar 2mm for the zoarium.

Fig. 48. Monticule-bearing bryozoan doubtfully referred to *Anomalotechus?* sp., Trepostomatida; encrusting a cephalopod; KCG collection; Lindwurm Member; collected by KST; scale bar 5mm.

Fig. 49. Ramose zoarium of *Fistulipora* sp., Cystoporata: Fistuliporidae; MPM P365916; Lindwurm Member; collected by KCG; scale bar 1cm.

Fig. 50. *Fenestella* sp., Fenestellida: Fenestellidae; MPM 5100; Berthelet Member; scale bar 5mm for the zoarium.

Fig. 51. Portion of a zoarium of *Fenestella* showing its characteristic net-like appearance, Fenestellida: Fenestellidae; UWM-GM 10048; Berthelet Member; scale bar 5mm.

Fig. 52. *Fenestella* sp., Fenestellida: Fenestellidae; UWM-GM 3511; Berthelet Member; scale bar 1cm.

Chapter 11 — Hederelloids

(Lophophorata: Phoronida?: Hederelloidea)

Hederelloids (Silurian–Permian) were originally classified as an order of encrusting cyclostome bryozoan, but have subsequently been removed from the Bryozoa because of their relatively large zooids, different budding patterns, and microprismatic walls, in contrast to the characteristics of the corresponding cyclostome features (Taylor & Wilson, 2008). Those authors suggested that hederelloids were instead colonial, phoronid-like invertebrates with retractable lophophores. Several hederelloid taxa have been found in the Milwaukee Formation and are shown here (Figs. 53–56). Cleland (1911) also listed *Hederella magna* Hall, 1881, although he considered it to be a bryozoan.

Fig. 53. The hederelloid *Hederella filiformis* (Billings, 1859); MPM 15676; Lindwurm Member; collected by GOR; scale bar 2mm.

Fig. 54. The hederelloid *Hederella cirrhosa* Hall, 1881; MPM 15692; Lindwurm Member; collected by GOR; scale bar 2mm.

Fig. 55. The hederelloid *Hernodia humifusa* Hall, 1881; MPM P28139; Berthelet Member; collected by KCG; scale bar 1cm.

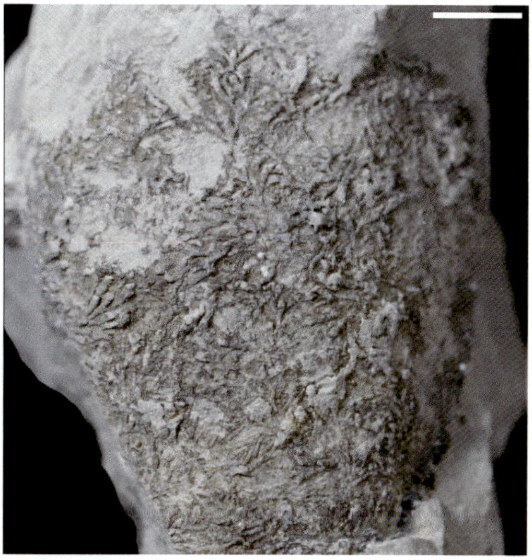

Fig. 56. Cephalopod encrusted by the hederelloid *Reptaria stolonifera* Rolle, 1851; MPM 15151; Lindwurm Member; collected by GOR; scale bar 1cm.

Chapter 12 — Brachiopods
(Lophophorata: Brachiopoda)

The Brachiopoda (Lower Cambrian–Holocene) is the lophophorate animal phylum consisting of solitary, marine, sessile, coelomate, metazoans having a pair of calcareous or chitinophosphatic valves that are bilaterally symmetrical about the median plane that is normal to the surface of separation between valves (Williams et al., 2000a) (Fig. 57). Molecular genetics and comparative anatomical studies suggest a close relationship between brachiopods, bryozoans, phoronids and other lophophore-bearing protostomes, which together comprise the Lophophorata. Like the Bryozoa, their placement in either the Deuterostomia or Protostomia is ambiguous on the basis of embryonic and molecular evidence.

As is typical of other Devonian marine faunas, the Brachiopoda is the Milwaukee Formation's most species-rich phylum, presently consisting of around forty-six named and un-named species (Griesemer, 1965) in three subphyla.

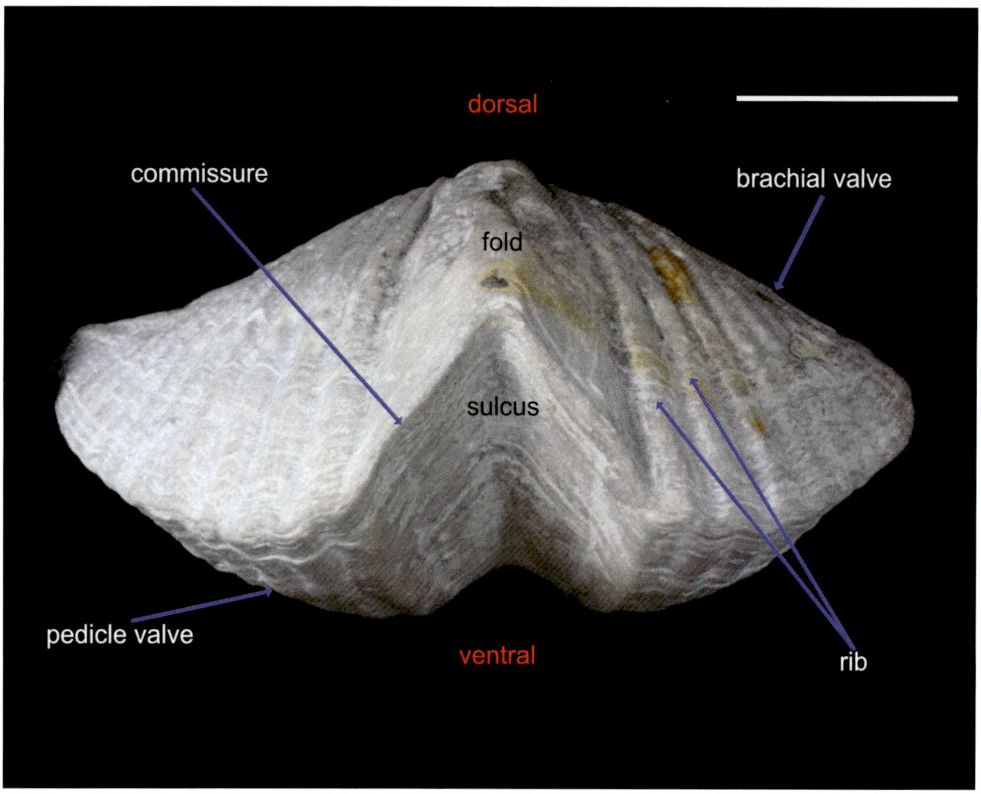

Fig. 57. Anterior view of the rhynchonelliform brachiopod *Orthospirifer milwaukeensis* (Cleland, 1911) with selected external morphological features and orientational terminology labeled; MPM 341; scale bar 1cm.

Linguliformea (Lower Cambrian–Holocene)

The linguliforms are brachiopods with organophosphatic stratiform shells that are not articulated or show rudimentary articulation not involving teeth and sockets. Living species possess a pedicle formed as an extension of the ventral body wall. It can be deduced that fossil forms did as well (Williams et al., 2000b).

Linguliform brachiopods are so named because of the general resemblance of many of its species to a tongue. They tend to deviate little from this basic shape and possess a rather conservative range and degree of ornamentation and size. Even their two valves can be difficult to distinguish. Their basic body plan was apparently very successful, and carrying on to the present day. Due to the apatitic composition of their shells producing a pale color (pink, purple or white) that contrasts with that of the surrounding matrix, linguliform brachiopods are easy to visually distinguish from other brachiopods.

Barroisella? milwaukeensis (Cleland, 1911), frequently preserves distinct color patterns (Fig. 58). That and other linguliform species are commonly preserved in the Milwaukee Formation (Figs. 59–65). In addition to the species shown here, Griesemer (1965) described the following taxa from at least one member of the Milwaukee Formation based on specimens that now reside at the Milwaukee Public Museum and/or Greene Geological Museum: *Barroisella?* cf. *complanata* (Williams, 1882), *Orbiculoidea* cf. *O. doria* (Hall, 1867), *Orbiculoidea* aff. *O. newberryi* (Hall, 1867), and *Orbiculoidea* sp. A.

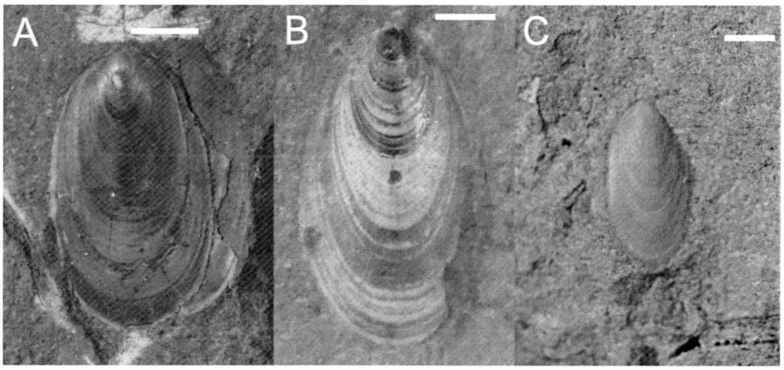

Fig. 58. *Barroisella? milwaukeensis* (Cleland, 1911), Lingulida: Lingulidae; apparent brachial valves. **A:** MPM 242; Berthelet Member; collected by CEM; scale bar 2mm. **B:** MPM 236; North Point Member; figured in Griesemer, 1965, Pl 1, Fig 4; collected by CEM; scale bar 1mm. **C:** MPM P28140; Berthelet Member; collected by KCG; scale bar 2mm.

Fig. 59. *Barroisella? rectilatera* (Hall, 1859), Lingulida: Lingulidae MPM 5136; Berthelet Member?; scale bar 5mm.

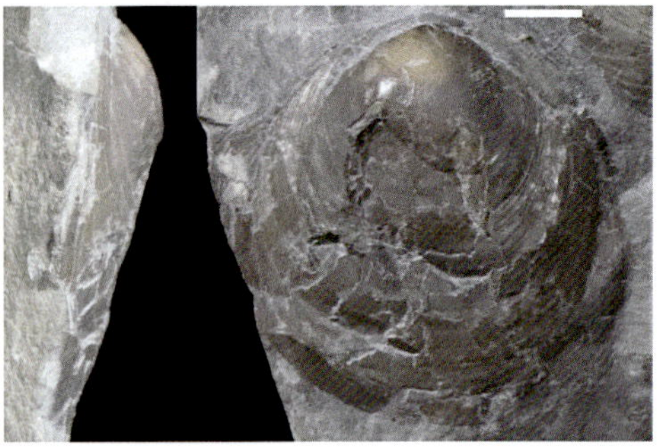

Fig. 60. Lateral and dorsal views of a brachial valve of *Lingulodiscina marginalis* (Whitfield, 1882), Lingulida: Discinidae; MPM 230; Berthelet Member; figured in Cleland, 1911, Pl 12, Figs 12&13; collected by CEM; scale bar 5mm.

Fig. 62. Brachial valve of *Trigonoglossa truncata* (Cleland, 1911), Lingulida: Obolidae; PAL 78258; holotype; Lindwurm Member; figured in Cleland, 1911, Pl 12, Fig 6; collected by EET; scale bar 5mm. Courtesy of Smithsonian Institution, Department of Paleobiology, National Museum of Natural History.

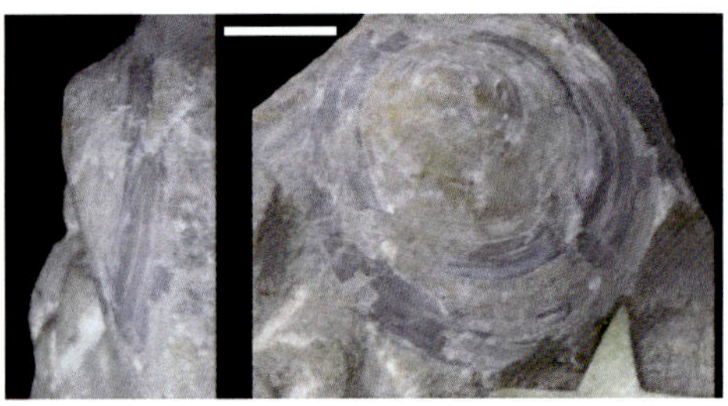

Fig. 64. Lateral and dorsal views of *Orbiculoidea telleri* Cleland, 1911, Lingulida: Discinidae; MPM 227; syntype; Lindwurm Member; figured in Cleland, 1911, Pl 12, Fig 9 and Griesemer, 1965, Pl 1, Figs 13&15; collected by CEM; scale bar 5mm

Fig. 61. A distorted strophodontid brachiopod mostly obscured by and hosting a crumbling *Lingulodiscina* sp.; MPM P28141; Lindwurm Member; collected by KCG; scale bar 5mm.

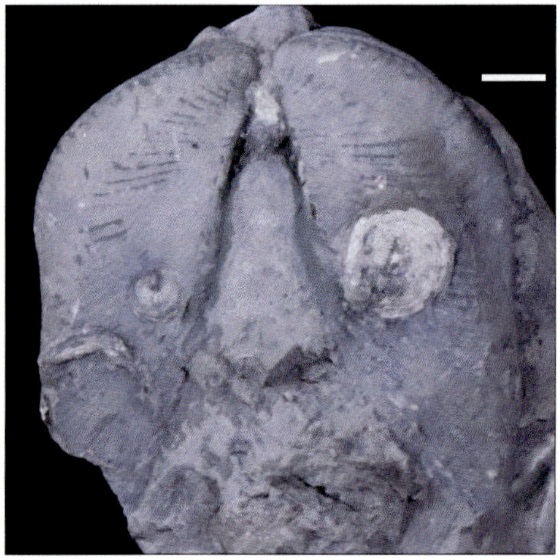

Fig. 63. Two linguliform brachiopods (*Orbiculoidea*) hosted by a splayed bivalve (*Grammysia*); UWM-GM 56200; Berthelet Member; scale bar 1cm.

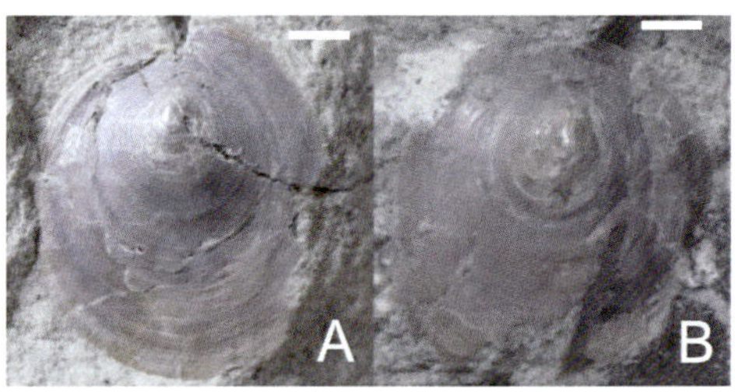

Fig. 65. *Orbiculoidea telleri* Cleland, 1911, Lingulida: Discinidae, dorsal; Berthelet Member; scale bars 2mm. **A:** MPM P28142; collected by KCG. **B:** MPM 225; figured in Cleland, 1911, Pl 12, Fig 10; collected by CEM.

Craniiformea (Lower Cambrian–Holocene)

Craniiform brachiopods are those with calcitic or aragonitic inarticulated shells with laminar (tabular) secondary layers. They lack a pedicle and either lie on the substrate or cement themselves by their ventral valves to other objects (Williams et al., 2000c).

Craniiforms are the least diverse of the brachiopods preserved in the Milwaukee Formation. The only species identified thus far is *Petrocrania hamiltoniae*? (Hall, 1867), which has only been found in the Berthelet Member and is represented by the interiors of the ventral valves cemented to larger shells (Figs. 66–68). Griesemer (1965, p. 255) tentatively assigned one particular specimen in the Greene Geological Museum to *Petrocrania* aff. *P. inflata* Cooper, 1956, although he suggests the possibility that it may be the dorsal valve of *P. hamiltoniae*.

Fig. 66. Interior of pedicle valve of *Petrocrania hamiltoniae?* (Hall, 1867) Craniida: Craniidae; MPM 233; Berthelet Member; figured in Cleland, 1911, Pl 12, Fig 8 and Griesemer, 1965, Pl 1, Fig 18; collected by CEM; scale bar 2mm.

Fig. 68. Brachial (top) and interior pedicle (bottom) valves of what appears to be *Petrocrania* sp., Craniida: Craniidae, hosted by a cephalopod; MPM P28144; Berthelet Member; collected by KCG; scale bar 5mm.

Fig. 67. Interior pedicle views of three specimens of *Petrocrania* sp., Craniida: Craniidae, hosted by a cephalopod; MPM P28143; Berthelet Member; collected by KCG; scale bar 5mm.

Rhynchonelliformea (Lower Cambrian–Holocene)

Rhynchonelliform brachiopods are those that have articulated calcitic shells with basically fibrous secondary layers (Williams et al., 2000d). Most species had a pedicle arising from the ventral valve with which they attached themselves to shells or other objects. They were the most common and diverse brachiopod subphylum in the Devonian, comprising over 75% of the brachiopod species in the Milwaukee Formation (Figs. 69–116). In addition to the species shown here, Griesemer (1965) discussed the following taxa from the Milwaukee Formation based on specimens that now reside at the Milwaukee Public Museum and/or Greene Geological Museum: *Pentamerella multicostata* Cleland, 1911, *Strophodonta* cf. *S. costata* Owen, 1852, *Strophodonta* sp. A, *Camarotoechia* cf. *C. prolifica* (Hall, 1867), *Elita* cf. *E. subundifera* (Meek & Worthen, 1868), *Cranaena* cf. *C. lincklaeni* (Hall, 1860), and *C. subovata* Savage, 1921.

Fig. 69. Dorsal, ventral, lateral, anterior, and posterior views of *Schizophoria* cf. *S. lata* Stainbrook, 1940, Orthida: Schizophoriidae; MPM 16436; Lindwurm Member; figured in Griesemer, 1965, Pl 2, Figs 5–7; collected by GOR; scale bar 5mm.

Fig. 70. Dorsal, ventral, lateral, and posterior views of *Schizophoria iowensis musculosa* Griesemer, 1965, Orthida: Schizophoriidae; MPM 16435; holotype; Lindwurm Member; figured in Griesemer, 1965, Pl 2, Figs 1&2; collected by GOR; scale bar 5mm.

Fig. 71. Dorsal and posterior views of in internal mold of *Schizophoria iowensis musculosa* Griesemer, 1965, Orthida: Schizophoriidae; MPM 15232; Berthelet Member; collected by GOR; scale bar 5mm.

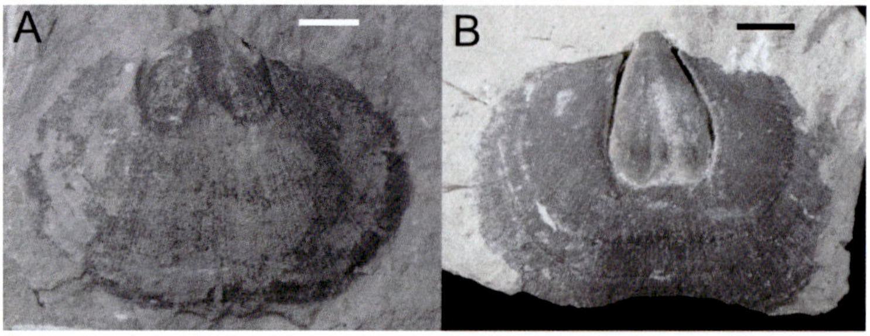

Fig. 72. *Schizophoria iowensis musculosa* Griesemer, 1965, Orthida: Schizophoriidae, Berthelet Member. **A:** Brachial, MPM 20445; figured in Griesemer, 1965, Pl 2, Fig 3, collected by SW. **B:** Pedicle, MPM 20446; figured in Griesemer, 1965, Pl 2, Fig 4; collected by SW. Scale bars 5mm.

Fig. 73. Dorsal, ventral, and lateral views of a small *Athyris vittata randalia* Stainbrook, 1942, Athyridida: Athyrididae; MPM P28146; Lindwurm Member; collected by KCG; scale bar 5mm.

Fig. 74. Dorsal, ventral, anterior, and posterior views of *Athyris vittata randalia* Stainbrook, 1942, Athyridida: Athyrididae; MPM 16602; Lindwurm Member; figured in Griesemer, 1965, Pl 6, Figs 1-3; collected by GOR; scale bar 5mm.

Fig. 75. Dorsal, ventral, anterior, and posterior views of *Tylothyris* cf. *T. bimesialis* (Hall, 1858), Spiriferinida: Mucrospiriferidae; MPM 359; Lindwurm Member; figured in Griesemer, 1965, Pl 5, Figs 13-15; scale bar is 5mm.

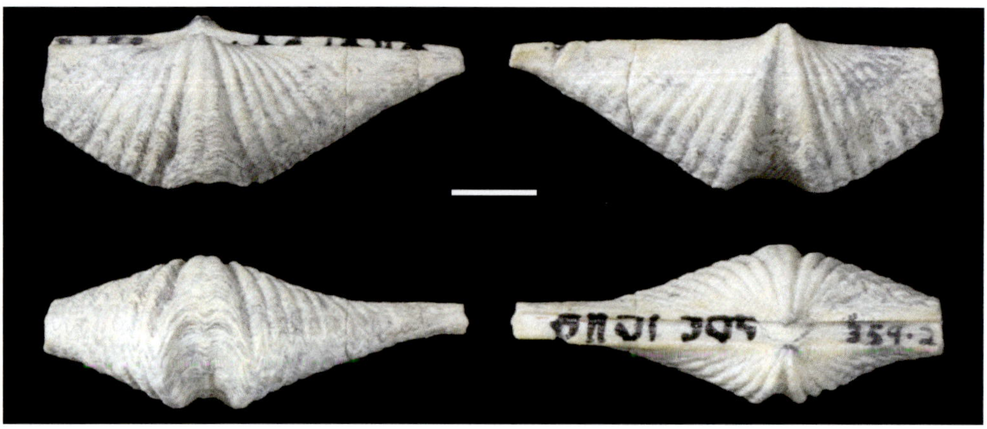

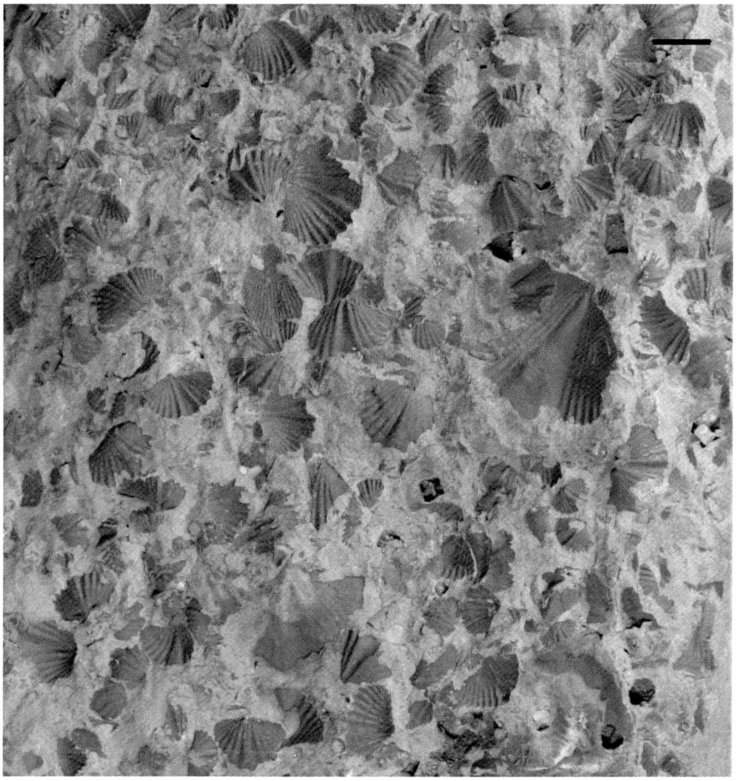

Fig. 76. Internal and external molds of brachiopod valves, mostly *Tylothyris subvaricosa umbonata* Griesemer, 1965; the largest mold appears to be *Orthospirifer milwaukeensis* (Cleland, 1911); UWM-GM 33112; Berthelet Member; scale bar 1cm.

Fig. 77. *Tylothyris subvaricosa umbonata* Griesemer, 1965, Spiriferinida: Mucrospiriferidae; MPM uncatalogued; Lindwurm Member; scale bar 5mm.

Fig. 78. Dorsal, ventral, lateral, anterior, and posterior views of *Mucrospirifer* sp., Spiriferinida: Mucrospiriferidae; MPM 358; Lindwurm Member; figured in Griesemer, 1965, Pl 5, Figs 1–2; collected by CEM; scale bar 5mm.

Fig. 79. Dorsal, ventral, lateral, anterior, and posterior views of *Orthospirifer milwaukeensis* (Cleland, 1911), Spiriferinida: Spinocyrtiidae; MPM 341; Lindwurm Member; figured in Cleland, 1911, Pl 15, Figs 1–5 and Griesemer, 1965, Pl 5, Figs 10–11; collected by CEM; scale bar 1cm.

Fig. 80. Dorsal, ventral, lateral, posterior, and close-up anterior views of *Orthospirifer clelandi* (Griesemer, 1965), Spiriferinida: Spinocyrtidae; MPM 16580; holotype; Lindwurm Member; figured in Griesemer, 1965, Pl 5, Figs 4–7; collected by GOR; scale bar 1cm for first 4 views and 2.5mm for close-up.

Fig. 81. Ventro-posterior view of *Orthospirifer iowensis* (Owen, 1852), Spiriferinida: Spinocyrtiidae; MPM 334; Lindwurm Member; figured in Cleland, 1911, Pl 17, Fig 4; collected by CEM; scale bar 1cm.

Fig. 82. Drawer full of internal molds of *Orthospirifer milwaukeensis* (Cleland, 1911), Spiriferinida: Spinocyrtiidae; UWM-GM individually numbered; Berthelet Member.

Fig. 83. Dorsal, lateral and posterior views of *Eosyringothyris* sp., Spiriferinida: Spinocyrtiidae; MPM 16507; Lindwurm Member; figured in Griesemer, 1965, Pl 4, Figs 4–6; collected by GOR; scale bar 5mm.

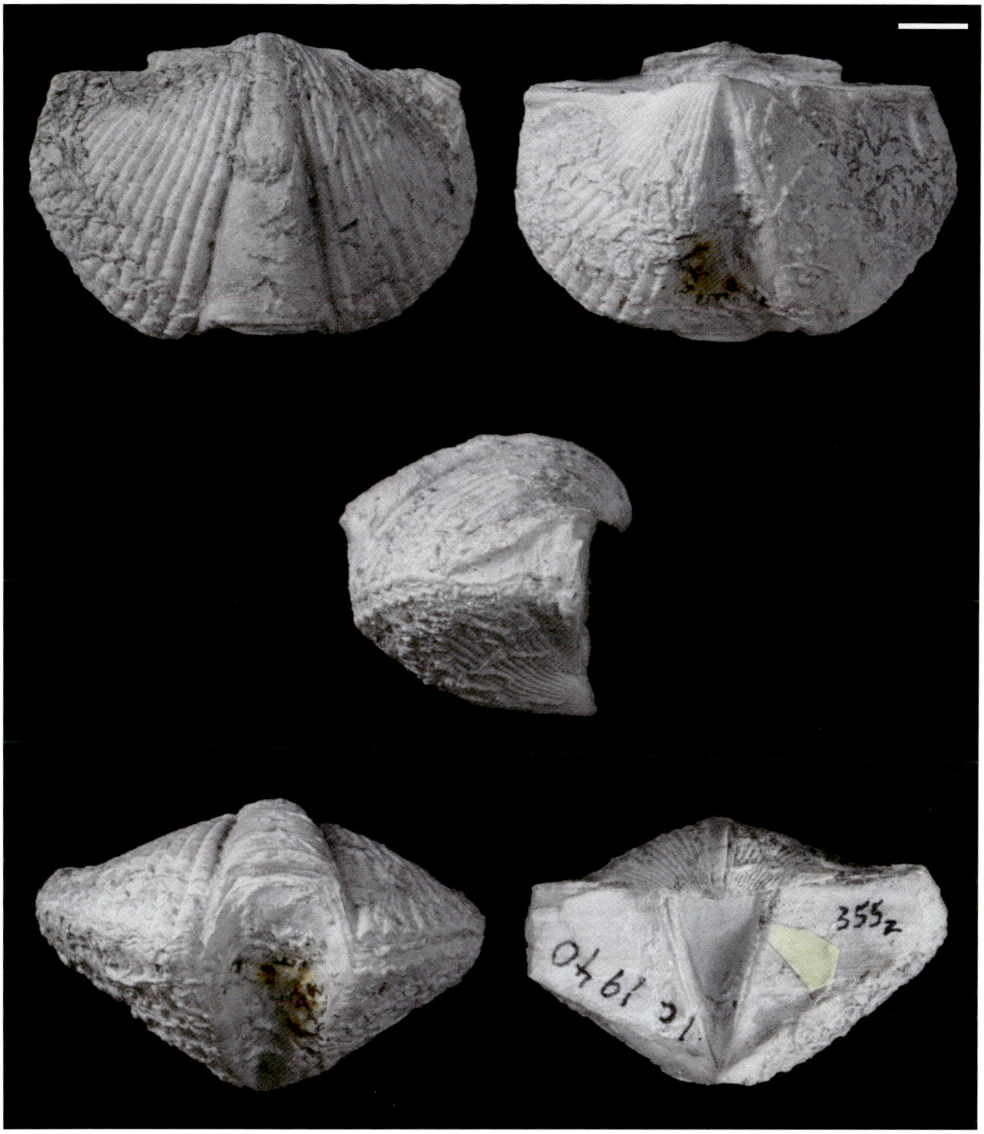

Fig. 84. Dorsal, ventral, lateral, anterior, and posterior views of *Eosyringothyris occidentalis* (Swallow, 1860), Spiriferinida: Spinocyrtiidae, hosting a reptant hederelloid; MPM 355; Lindwurm Member; figured in Griesemer, 1965, Pl 4, Figs 19–21; collected by CEM; scale bar 5mm.

Fig. 85. Dorsal, ventral, and lateral views of *Cyrtina umbonata* (Hall, 1858), Spiriferinida: Cyrtinidae; MPM 16491; Lindwurm Member; figured in Griesemer, 1965, Pl 6, Figs 10&11; collected by GOR; scale bar 5mm.

Fig. 86. *Cyrtina triquetra* (Hall, 1858), Spiriferinida: Cyrtinidae; MPM uncatalogued; Lindwurm Member; scale bar 5mm.

Fig. 87. *Cyrtina umbonata* (Hall, 1858), Spiriferinida: Cyrtinidae MPM uncatalogued; Lindwurm Member; scale bar 5mm. Notice that the pedicle valve projects well beyond the brachial valve toward the posterior. This is not the case with *C. triquetra*.

Fig. 88. Ventral and lateral views of *Cyrtina triquetra* (Hall, 1858), Spiriferinida: Cyrtinidae; MPM 365; Lindwurm Member; figured in Griesemer, 1965, Pl 6, Figs 7–9; collected by CEM; scale bar 2mm.

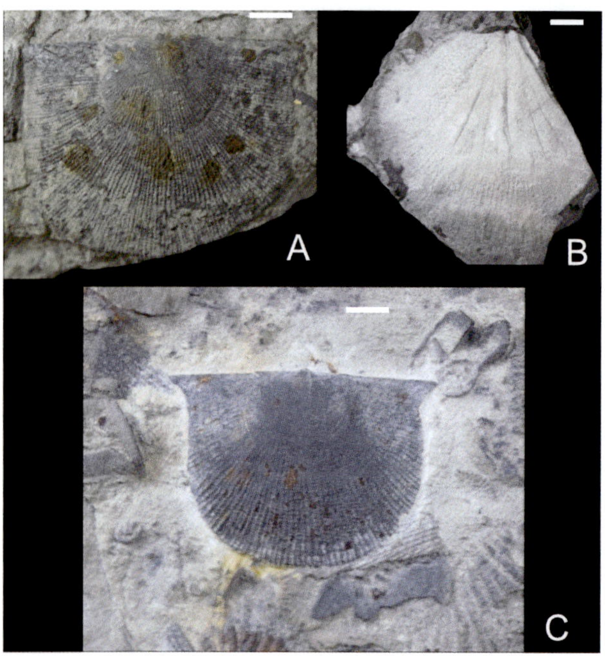

Fig. 89. *Protoleptostrophia serrata* Griesemer, 1965; Strophomenida: Leptostrophiidae; Lindwurm Member; scale bar is 5mm. **A:** Brachial; MPM 317; collected by CEM. **B:** Pedicle interior; MPM 321; figured in Griesemer, 1965, Pl 2, Fig 16. **C:** Brachial; UWM-GM G2073.

Fig. 90. Interior of brachial valve of *Protoleptostrophia serrata* Griesemer, 1965, Strophomenida: Leptostrophiidae; MPM P28196; Lindwurm Member; figured in Griesemer, 1965, Pl 2, Fig 17; scale bar 5mm.

Fig. 91. *Pholidostrophia (Pholidostrophia) iowensis obscura* Griesemer, 1965, Strophomenida: Strophodontidae; brachial valve, MPM 5021; holotype; Lindwurm Member; figured in Griesemer, 1965, Pl 2, Figs 13–15; scale bar 5mm.

Fig. 92. Interior of brachial valve of *Pholidostrophia (Pholidostrophia) iowensis obscura* Griesemer, 1965, Strophomenida: Strophodontidae, encrusted by a reptant hederelloid; MPM 16352; Lindwurm Member; figured in Griesemer, 1965, Pl 2, Fig 13; collected by GOR; scale bar 2mm.

Fig. 93. Brachial valve of *Strophodonta halli* Cleland, 1911, hosting the coral, *Aulopora*; MPM P28744 B; Lindwurm Member; collected by KCG; scale bar 1cm; prepared by David Comfort.

Fig. 94. Slab of *Strophodonta halli* Cleland, 1911; Strophomenida: Strophodontidae; UWM-GM 33112; Berthelet Member; scale bar 1cm.

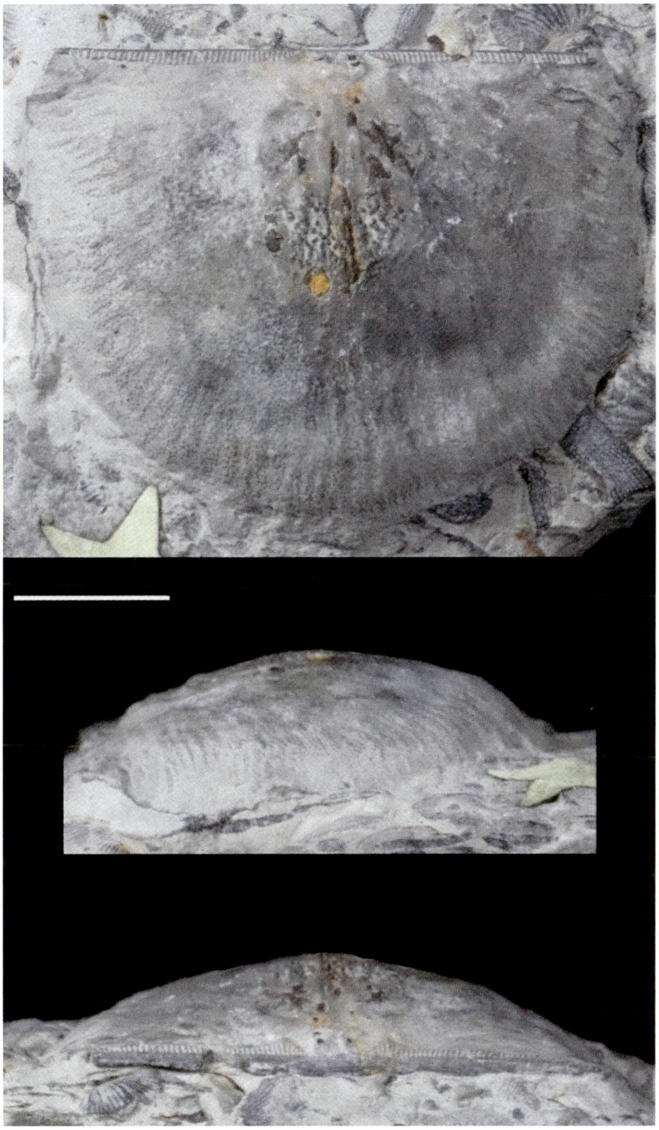

Fig. 95. Interior dorsal, lateral and posterior of brachial valve of *Strophodonta halli* Cleland, 1911, Strophomenida: Strophodontidae; MPM 310; Lindwurm Member; figured in Cleland, 1911, Pl 14, Fig 8 and Griesemer, 1965, Pl 2, Fig 23; collected by CEM; scale bar 1cm.

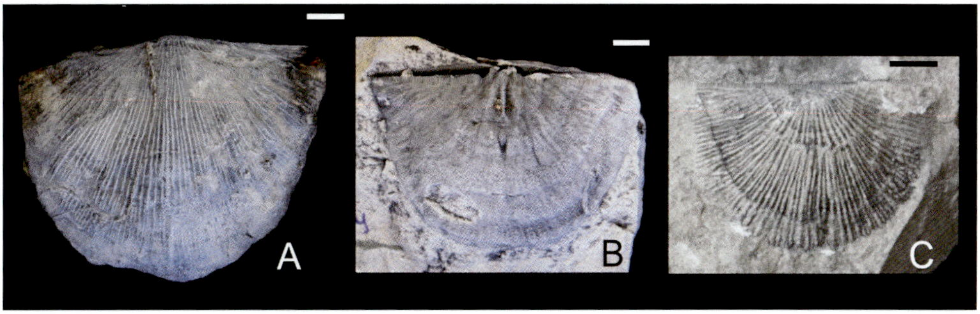

Fig. 96. *Strophodonta halli* Cleland, 1911, Strophomenida: Strophodontidae; scale bar 5mm. **A:** Brachial valve; PAL 111811; syntype, Lindwurm Member; figured in Cleland, 1911, Pl 14, Fig 7; collected by EET; courtesy of Smithsonian Institution, Department of Paleobiology, National Museum of Natural History. **B:** Interior of brachial valve; PAL 85972; syntype; Berthelet Member; figured in Cleland, 1911, Pl 14, Fig 10; collected by EET; courtesy of Smithsonian Institution, Department of Paleobiology, National Museum of Natural History. **C:** Brachial valve; MPM 20439; Berthelet Member; collected by SW.

Fig. 97. *Floweria? arctostriata* (Hall, 1843), Orthotetida: Areostrophiidae; Lindwurm Member. **A:** Dorsal; MPM 16364; collected by GOR; scale bar 2mm; **B:** Ventral; MPM 1843; figured in Griesemer, 1965, Pl 3, Fig 5; scale bar 5mm.

Fig. 98. Mass association of *Striatochonetes schucherti* (Cleland, 1911), Chonetida: Chonetidae; other brachiopods, crinoid columnals, and bryozoans are also present; MPM P28197; Lindwurm Member; scale bar 1cm.

Fig. 99. An assortment of brachiopods, mostly *Striatochonetes schucherti* (Cleland, 1911), Chonetida: Chonetidae; MPM P28147; Lindwurm Member; collected by KCG; scale bar 5mm.

Fig. 100. Three brachial valves of *Longispina?* sp, Chonetida: Chonetidae; MPM P28189; Berthelet Member; collected by KCG; scale bar 5mm.

Fig. 101. *Longispina?* sp., Chonetida: Chonetidae; MPM 5027; Lindwurm Member; scale bar 2mm.

Fig. 102. Dorsal, ventral, lateral, posterior, and anterior views of a large specimen of *Devonatrypa?* sp. A, Atrypida: Atrypidae; MPM P16461; Lindwurm Member; collected by GOR; scale bar 1cm.

Fig. 103. Dorsal view of two specimens of *Devonatrypa?* sp. B, Atrypida: Atrypidae; Lindwurm Member; collected by KCG; scale bar 5mm. **A:** MPM P28148. **B:** This specimen is hosting a commensal, possibly an edrioasteroid; MPM P28149.

Fig. 104. Slab with numerous brachiopods, mostly *Spinatrypa spinosa milwaukeensis* Griesemer, 1965, Atrypida: Atrypidae; MPM P28190; Berthelet Member, "*Atrypa hystrix* band;" scale bar 1cm.

Fig. 105. *Spinatrypa spinosa milwaukeensis* Griesemer, 1965, Atrypida: Atrypidae; Berthelet Member; scale bars 5mm. **A:** Dorsal valve; MPM P28150; collected by KCG; arrow points to marginal spine. **B:** Dorsal; MPM 20442; holotype; figured in Griesemer, 1965, Pl 4, Fig 16; collected by SW. **C:** Dorsal of internal mold on left; dorsal of external mold on right; MPM 15251; figured in Griesemer, 1965, Pl 4, Fig 17 on left, Fig 15 on right; collected by GOR.

Fig. 106. Dorsal, ventral, lateral, and posterior views of *Desquamatia (Seratrypa)?* sp., Atrypida: Atrypidae; MPM 16464; Lindwurm Member; scale bar 5mm.

Fig. 107. Dorsal, ventral, lateral, anterior, and posterior views of *Desquamatia (Seratrypa)?* sp., Atrypida: Atrypidae; MPM 19792; Lindwurm Member; figured in Griesemer, 1965, Pl 4, Figs 5–7; collected by BB; scale bar 5mm.

Fig. 108. Ventral, dorsal, lateral, and posterior views of small specimen of *Desquamatia (Seratrypa)?* sp., Atrypida: Atrypidae; MPM 16473; Lindwurm Member; figured in Griesemer, 1965, Pl 4, Fig 8; collected by GOR; scale bar 2mm.

Fig. 109. Six specimens of *Eumetabolotoechia greeni* (Cleland, 1911), Rhynchonellida: Eumetabolotoechiidae; MPM-KCG uncatalogued; collected by KCG; scale bar 1cm. Originally assigned to *Leiorhynchus*, this species was reassigned to *Eumetabolotoechia* by Sartenaer (2014).

Fig. 110. *Cupularostrum depressa* (Kindle, 1901), Rhynchonellida: Trigonirhynchidae. **A:** Dorsal view, MPM 378; Lindwurm Member; figured in Griesemer, 1965, Pl 3, Fig 17; collected by CEM; scale bar 2mm. **B:** Ventral?, MPM P28145; Berthelet Member; collected by KCG; scale bar 5mm.

Fig. 111. Dorsal, lateral, anterior, and posterior views of *Cupularostrum scitulus* (Cleland, 1911), Rhynchonellida: Trigonirhynchiidae; UWM-GM 5162; Berthelet Member; figured in Griesemer, 1965, Pl 3, Figs 24–26; scale bar 2mm.

Fig. 112. Dorsal, ventral, and lateral views of *Cryptonella?* aff. *C. reimanni* Cloud, 1942, Terebratulida: Cryptonellidae; MPM 16491; Lindwurm Member; figured in Griesemer, 1965, Pl 6, Figs 12&13; collected by GOR; scale bar 5mm.

Fig. 113. Dorsal, ventral, and lateral views of what appears to be *Cranaena thomasi* Stainbrook, 1941, Terebratulida: Cranaenidae; MPM P28726; collected by JLG; scale bar 5mm.

Fig. 114. Dorsal, ventral, lateral, posterior, and anterior views of *Cranaena* cf. *C. cooperi* Cloud, 1941, Terebratulida: Cranaenidae; WESM-KCG 1129; Lindwurm Member; collected by TCG; scale bar 5mm; prepared by David Comfort.

Fig. 115. Six cranaenid brachiopod valves showing color patterns; Lindwurm Member; **A:** Brachial valve; MPM P28150; collected by KCG; scale bar 5mm. **B:** Brachial valve; MPM 16460; collected by GOR; scale bar 2mm. **C:** Pedicle valve; MPM 381; collected by CEM; scale bar 5mm. **D:** Splayed valves; WESM-KCG 1115; collected by LVS; scale bar 5mm. **E:** Brachial valve; MPM 19798; collected by BB; scale bar 5mm.

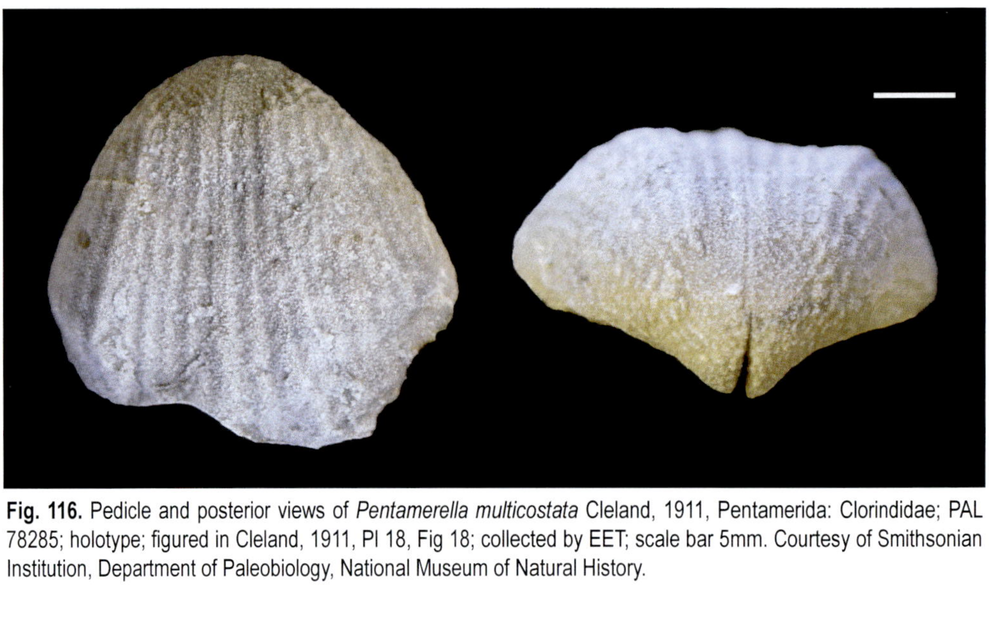

Fig. 116. Pedicle and posterior views of *Pentamerella multicostata* Cleland, 1911, Pentamerida: Clorindidae; PAL 78285; holotype; figured in Cleland, 1911, Pl 18, Fig 18; collected by EET; scale bar 5mm. Courtesy of Smithsonian Institution, Department of Paleobiology, National Museum of Natural History.

Chapter 13 — Hyoliths
(Lophophorata?: Hyolithida)

Hyoliths (Cambrian–Permian) are known mainly from fossils of their calcareous, bilaterally symmetrical, conoidal sclerites that are open at their wider ends. The other hyolith sclerites, consisting of an operculum and a pair of lateral spines (helens), are sometimes preserved as well. Historically regarded as akin to the mollusks or placed in their own phylum, hyoliths have recently been compared to the lophophorates, based on well preserved hyoliths from the Burgess and Spence Shale Lagerstätten (Moysiuk et al., 2017).

A few conoidal hyolithid sclerites have turned up in the Berthelet Member (Fig. 117). Cleland (1911) tentatively attributed some of them to *Hyolithes aclis*? Hall, 1876.

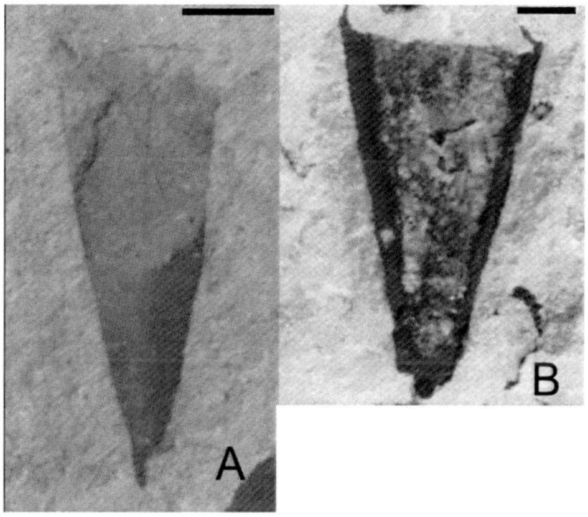

Fig. 117. *Hyolithes* sp., Hyolithida: Hyolithidae, Berthelet Member; scale bar 2mm. **A:** MPM 272; figured in Cleland, 1911, Pl 26, Fig 8; collected by CEM. **B:** MPM P28151; collected by KCG.

Chapter 14 — Mollusks
(Mollusca)

The Mollusca (Cambrian–Holocene) is the protostome animal phylum consisting of aqueous or terrestrial nonsegmented (except for the Polyplacophora) invertebrates having a fleshy mantle that secretes a calcareous shell of one or more pieces. The shell may be exteriorly-facing, enclosed by the mantle, reduced or lost (Yonge, 1960). It is an important and diverse phylum whose classes include the Monoplacophora, Polyplacophora, Gastropoda, Rostroconchia, Bivalvia, Scaphopoda, Cephalopoda, and several living and extinct classes of lesser importance. The mollusk classes found in the Milwaukee Formation are represented by the gastropods, rostroconchs, bivalves and cephalopods.

Gastropoda (Cambrian–Holocene)

The Gastropoda is a diverse class of mollusks that includes snails and slugs and typically has a spiral calcareous shell, although some shells are conical, cap-shaped or absent (Fig. 118). A fleshy foot typically protrudes from the shell and serves for ambulation. Most gastropods are vagile benthos, inhabiting marine or freshwater environments, while sessile or terrestrial taxa exist as well. A keratinaceous radula is used for feeding and, in some forms, for boring through living shells to feed on the internal tissues of its prey. Many forms also possess a keratinous operculum, which is used to close off the aperture of the shell (Cox, 1960).

The Milwaukee Formation's gastropod shells include both low and high spired forms. The most common gastropods in this fauna, the platyceratids, cemented themselves to other objects, mainly crinoid hosts, shaping their apertures to conform with the outline of the surface to which they were attached.

Several of the gastropod species from the Milwaukee Formation are illustrated here (Figs. 119–132). Cleland (1911) included the following additional species: *Pleurotomaria* sp., *Porcellia*? *kindlei* Cleland, 1911, *Bellerophon (Ptomatis) patulus* Hall, 1879, and *Platyceras*? *bertheletensis unsymmetricum* Cleland, 1911.

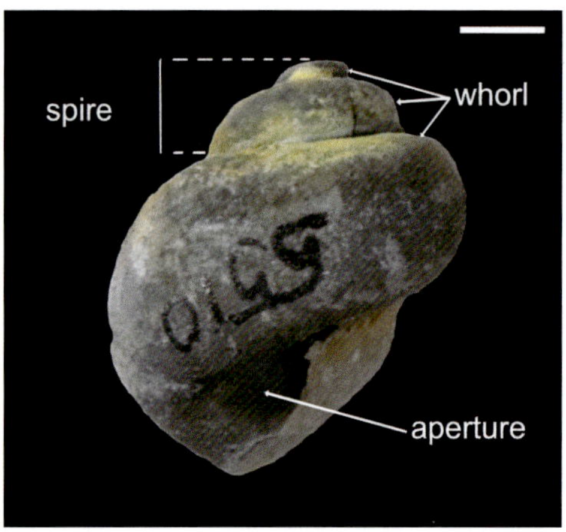

Fig. 118. Internal mold of the gastropod *Naticonema?* sp. in standard orientation (i.e., spire pointing upwards and aperture facing observer) with selected external morphological features labeled; UWM-GM 5510; scale bar 5mm.

Fig. 119. Two views of *Platyceras? bertheletensis* Cleland, 1911, Archaeogastropoda: Platyceratidae; PAL 78211; holotype; Berthelet Member; figured in Cleland, 1911, Pl 28, Fig 2; collected by EET; scale bar 5mm. Courtesy of Smithsonian Institution, Department of Paleobiology, National Museum of Natural History.

Fig. 120. Three views of *Platyceras? hornefferi* Cleland, 1911, Archaeogastropoda: Platyceratidae; MPM 393; syntype; Lindwurm Member; figured in Cleland, 1911, Pl 28, Figs 4&5; collected by CEM; scale bar 5mm.

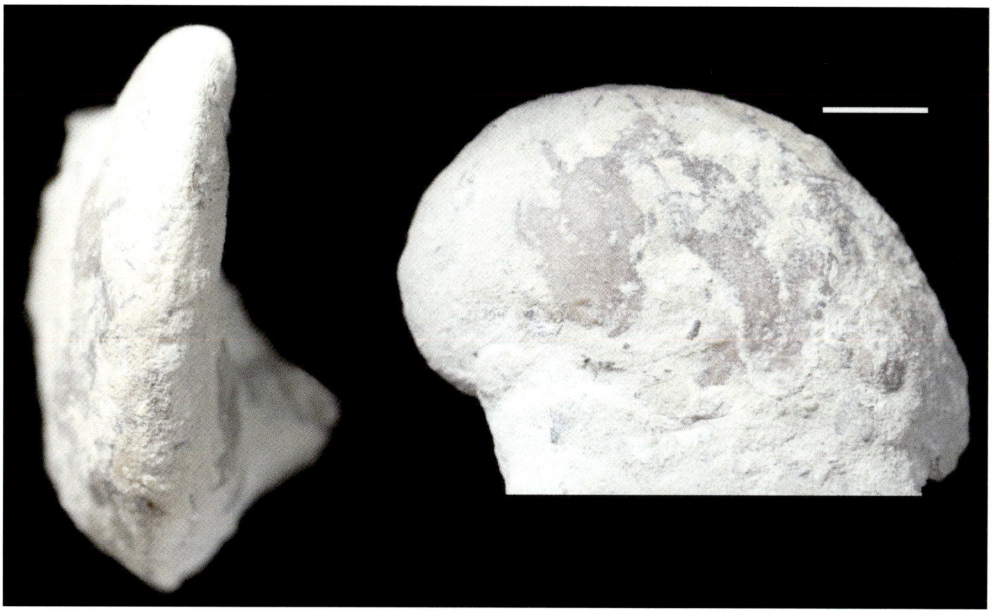

Fig. 121. Two views of *Platyceras?* sp., Archaeogastropoda: Platyceratidae; MPM P28152; Lindwurm Member; collected by KCG; scale bar 5mm.

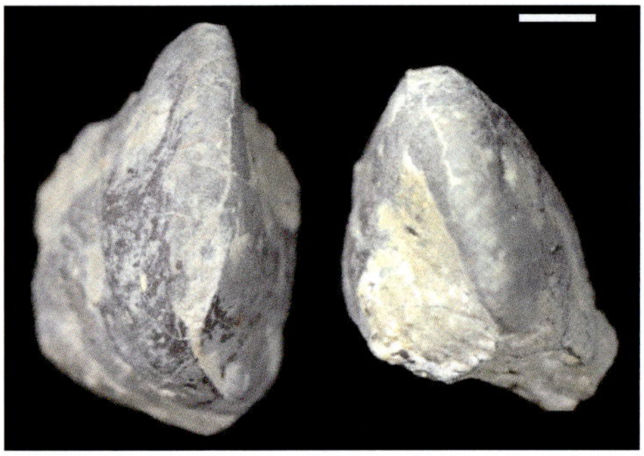

Fig. 122. Two views of a testiferous specimen of *Platyceras?* sp. Archaeogastropoda: Platyceratidae; MPM P28153; Lindwurm Member; collected by KCG; scale bar 5mm.

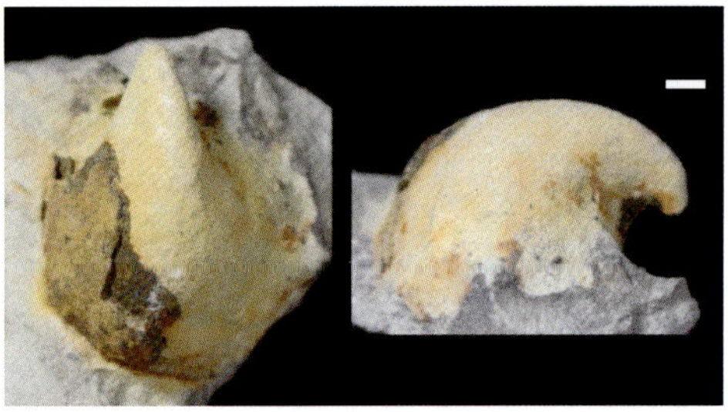

Fig. 124. Two views of an internal mold of *Platyceras?* sp., Archaeogastropoda: Platyceratidae; MPM P28154; Lindwurm Member; collected by KCG; scale bar 2mm.

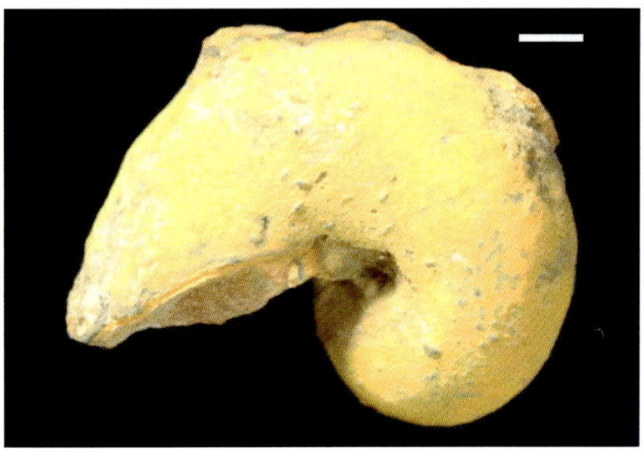

Fig. 126. *Naticonema?* sp., Archaeogastropoda: Platyceratidae; MPM P28155; Lindwurm Member; collected by KCG; scale bar 2mm.

Fig. 128. Two views of a silicified specimen of *Cyclonema? subglobosa* Cleland, 1911, Archaeogastropoda: Platyceratidae; MPM 398; North Point Member; North Avenue Intake Tunnel; figured in Cleland, 1911, Pl 28, Fig 7; collected by CEM; scale bar 5mm.

Fig. 123. An assortment of platyceratids, most with their marcasite shells weathered and crumbling away to varying degrees; MPM-KCG uncatalogued; Lindwurm Member; collected by KCG; scale bar 1cm.

Fig. 125. *Naticonema?* sp., Archaeogastropoda: Platyceratidae; UWM-GM 5510; Lindwurm Member; scale bar 5mm.

Fig. 127. An assortment of *Naticonema?* shells, Archaeogastropoda: Platyceratidae; UWM-GM 5430, 5388 etc; Lindwurm Member; scale bar is 5mm.

Fig. 129. Silicified specimen of *Bellerophon pelops* Hall, 1879, Archaeogastropoda: Bellerophontidae; MPM 395; North Point Member; North Avenue Intake Tunnel; figured in Cleland, 1911, Pl 28, Fig 12; collected by CEM; scale bar 5mm.

Fig. 130. *Phanerotinus* sp., Archaeogastropoda: Euomphalidae; MPM P28156; Berthelet Member; collected by KCG; scale bar 2mm.

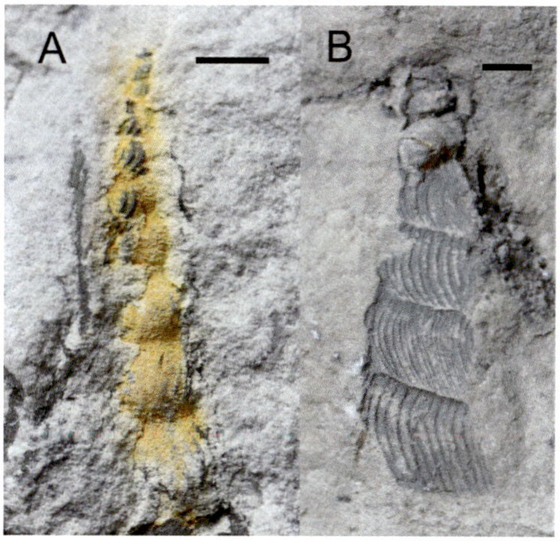

Fig. 132. *Loxonema hamiltoniae* Hall, 1862, Caenogastropoda: Loxonomatidae; collected by KCG. **A:** MPM P28157; Lindwurm Member; scale bar 2mm. **B:** MPM P28158; Berthelet Member; scale bar 5mm.

Fig. 131. External mold of *Loxonema hamiltoniae* Hall, 1862, Caenogastropoda: Loxonomatidae; MPM 396; North Point Member; North Avenue Intake Tunnel; figured in Cleland, 1911, Pl 27, Figs 5&6; collected by CEM; scale bar 2mm.

Rostroconchia (Cambrian–Permian)
This was an odd Paleozoic class of mollusks probably including the ancestors of bivalves and with a bilaterally symmetrical shell and a "pseudohinge." In other words, the hinge area was not fully decalcified, even though the animal apparently lived in a burrow. They literally had to crack the skeletal material in the dorsal hinge line when the shells grew larger. One species of rostroconch is currently known from the Milwaukee Formation (Fig. 133).

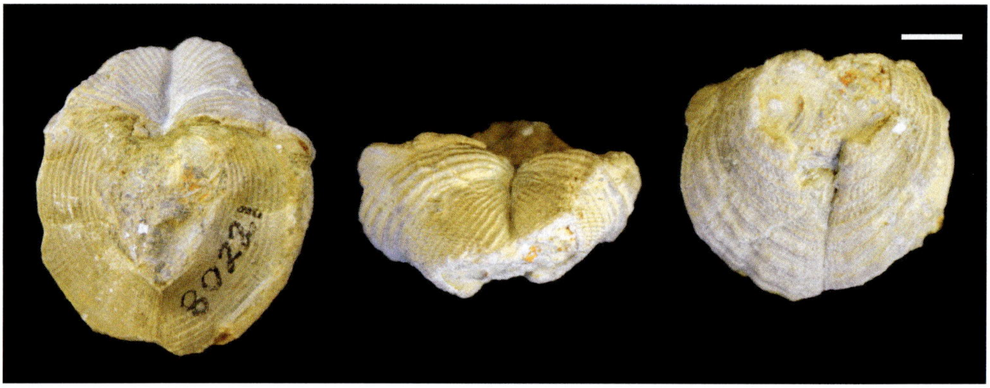

Fig. 133. Three views of *Conocardium ornatum* Cleland, 1911, Conocardioida: Conocardiidae; PAL 80221; holotype; Lindwurm Member; figured in Cleland, 1911, Pl 21, Figs 9&10 and Pohl, 1929, Pl 7, Figs 2–4; collected by EET; scale bar 5mm. Courtesy of Smithsonian Institution, Department of Paleobiology, National Museum of Natural History.

Bivalvia (Cambrian–Holocene)
The Bivalvia is the class of aquatic mollusks that includes the clams, oysters, mussels and related forms typically possessing two calcareous valves that hinge together and enclose a muscular foot and other soft parts. Unlike with the brachiopods, bivalve shells are not bilaterally symmetrical. Instead, the plane of symmetry lies along the surface where the two shells come into contact (hingeline)—the shells being mirror images of each other (Cox, 1969) (Fig. 134). Bivalves are typically infaunal, burrowing with their foot into the sediment beneath bodies of water, while others are vagile or sessile benthos, inhabiting a wide variety of aquatic environments. A few forms, such as certain scallops, are capable of swimming by repeatedly opening and snapping their shells closed.

The Milwaukee Formation has a diverse fauna of bivalves, including 43 named species or subspecies of clams, scallops, mussels and other forms, 30 of which are figured here (Figs. 135–182). The small clam *Palaeoneilo* was the most abundant bivalve. The fossils of its many species together easily make up more than half of all Milwaukee Formation bivalves combined.

It is interesting to note the preservation of many of the bivalves in this formation. The valves in certain layers are typically preserved in such a way as to show to varying degrees both internal and external features, whereby for example what appears to be an internal mold often exhibits distinct growth lines. This represents in one case a charcoal gray residue of the original shell material, which is probably a carbonized remnant of the periostracum. In other cases the remnants of the shell consist of an orange material such as marcasite that apparently replaced the original shell (see Chapter 4), and then variably weathered from the higher regions of the fossil. In both cases the result is a distint outline of the shell's growth lines and margins. When the shell is completely weathered away, interior features become visible. The coloration tends to make the outline of the fossils show up better in photographs in comparison with that of many fully testiferous or permineralized fossils. It can be observed on many of the bivalves and certain other calcareous skeletal materials shown in this book.

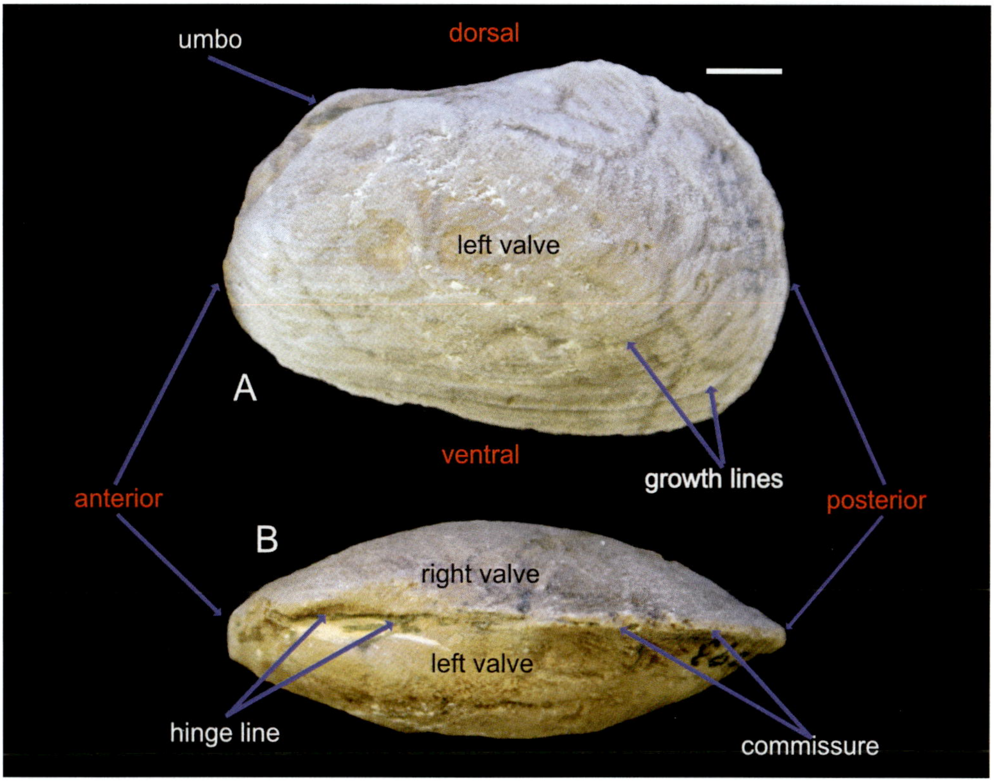

Fig. 134. The bivalve *Modiomorpha mytiloides milwaukeensis* Pohl, 1929 with selected external morphological features and orientational terminology labeled; PAL 80293; scale bar 5mm. Courtesy of Smithsonian Institution, Department of Paleobiology, National Museum of Natural History. **A:** Full view of left valve. **B:** Dorsal view.

E.R. Pohl (1929) established many bivalve species based on material from the Milwaukee Formation, many of which are known only from one or two fragmentary, incompletely prepared specimens. In addition to the species figured below, Pohl described four un-named species and named and/or described the following named species from the Milwaukee Formation based on specimens that now reside at the Milwaukee Public Museum or the National Museum of Natural History: *Grammysia regularis* Pohl, 1929, *Cimataria obtusiloba* (Cleland, 1911), *Nucula triangula* Pohl, 1929, *Palaeoneilo corrugata angulata* Pohl, 1929, *P. persinuata* Pohl, 1929, *Nuculana gibbosa* Pohl, 1929, *Megambonia wisconsinensis* Cleland, 1911, *Lophonychia cultellata* Pohl, 1929, *Aviculopecten bassleri* Pohl, 1929, *Aviculopecten hystriculus* Pohl, 1929, *Pterinopecten (Vertumnia) simplex* (Pohl, 1929), and *Modiomorpha pediformis* Pohl, 1929. A modern analysis of the Milwaukee Formation's bivalves is needed.

Fig. 135. Splayed interiors of *Modiomorpha saccula* Pohl, 1929), Modiomorphoida: Modiomorphidae; MPM P28159; Lindwurm Member; collected by KCG; scale bar 5mm.

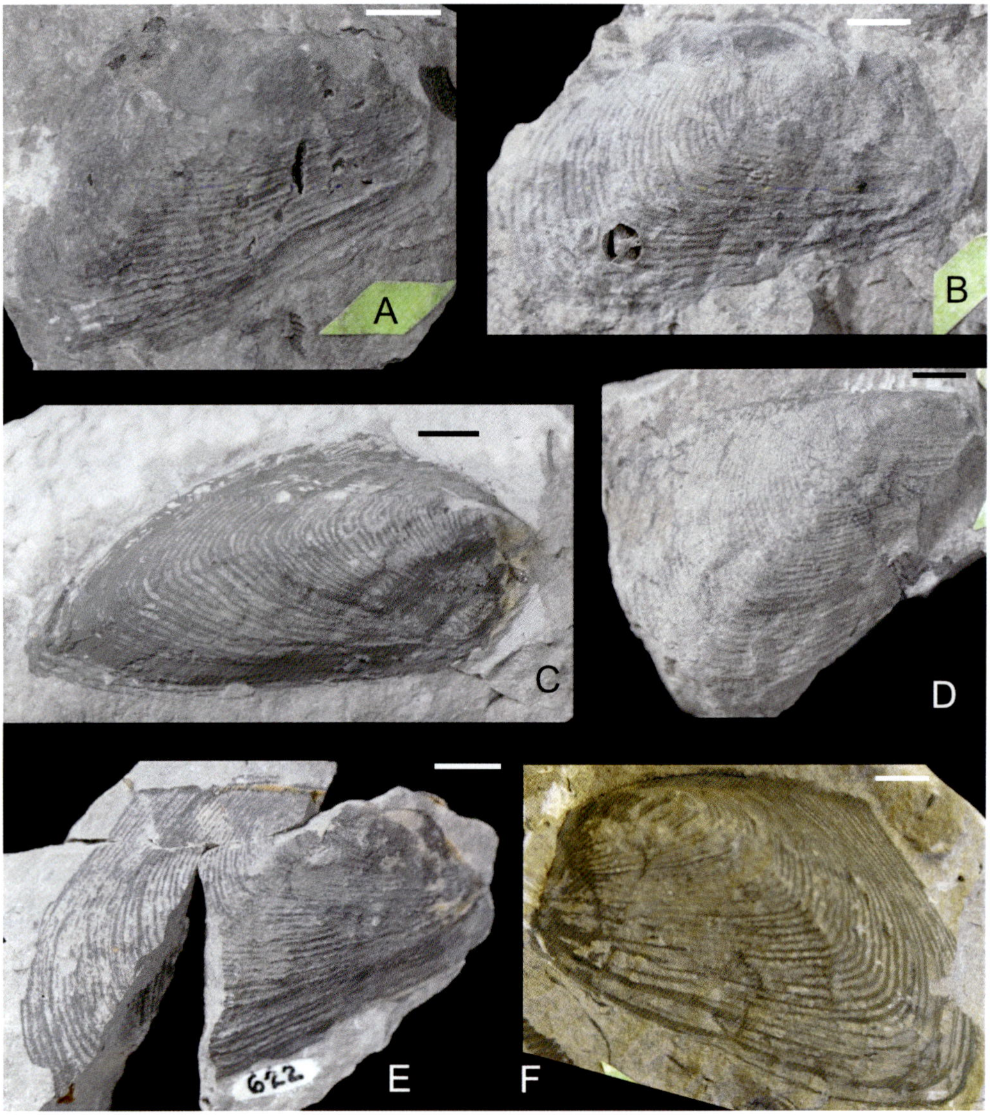

Fig. 136. *Modiomorpha saccula* Pohl, 1929, Modiomorphoida: Modiomorphidae; Berthelet Member; scale bar 5mm. **A:** Right valve; MPM 27637; syntype; figured in Pohl, 1929, Pl 11, Fig 2. **B:** Right valve; MPM 2451; syntype; figured in Pohl, 1929, Pl 11, Fig 3. **C:** Right valve; MPM 13575; syntype; figured in Pohl, 1929, Pl 10, Fig 12; collected by GOR. **D:** Right valve; MPM 458; syntype; figured in Pohl, 1929, Pl 10, Fig 14; collected by CEM. **E:** Right valve; MPM P28160; collected by KCG; a well outlined specimen which may belong to this species. **F:** Left valve; PAL 80312; syntype; figured in Pohl, 1929, Pl 10, Fig 9; Courtesy of Smithsonian Institution, Department of Paleobiology, National Museum of Natural History.

Fig. 137. Lateral and full view of left valve of *Modiomorpha schucherti* Cleland, 1911, Modiomorphoida: Modiomorphidae; MPM 495; syntype; Berthelet Member; figured in Cleland, 1911, Pl 24, Fig 3 and Pohl, 1929, Pl 13, Fig 6; collected by CEM; scale bar 5mm.

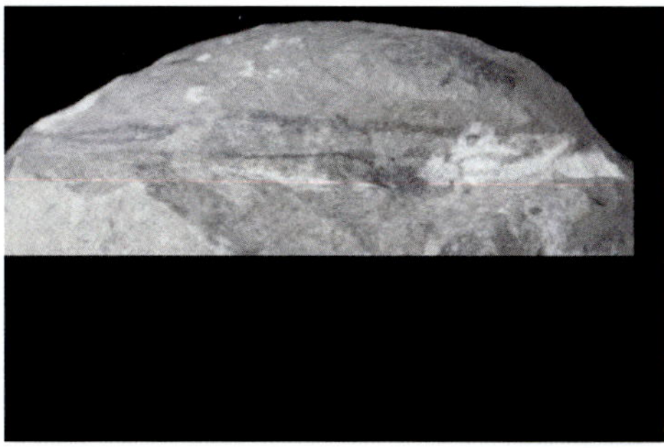

Fig. 138. Left valve of *Modiomorpha diffidens* Pohl, 1929, Modiomorphoida: Modiomorphidae; MPM 13569; holotype; Berthelet Member; figured in Pohl, 1929, Pl 14, Fig 5; collected by GOR; scale bar 2mm.

Fig. 140. Left valve of *Modiomorpha schucherti* Cleland, 1911, Modiomorphoida: Modiomorphidae; MPM P28161; Berthelet Member; collected by KCG; scale bar 5mm.

Fig. 141. Right valve of *Modiomorpha elongata* (Cleland, 1911), Modiomorphoida: Modiomorphidae; MPM 18599; Berthelet Member; figured in Pohl, 1929, Pl 1, Fig 6; collected by TAG; scale bar 5mm.

Fig. 139. Left valve and dorsal view of *Modiomorpha mytiloides milwaukeensis* Pohl, 1929, Modiomorphoida: Modiomorphidae; PAL 80293; holotype; Berthelet Member; figured in Pohl, 1929, Pl 12, Figs 4&5; collected by EET; scale bar 5mm. Courtesy of Smithsonian Institution, Department of Paleobiology, National Museum of Natural History.

Fig. 142. Dorsal and full left valve views of *Modiomorpha elongata* (Cleland, 1911), Modiomorphoida: Modiomorphidae; MPM 460; holotype; Berthelet Member; figured in Cleland, 1911, Pl 22, Figs 10–12 and Pohl, 1929, Pl 11, Fig 5; collected by CEM; scale bar 5mm.

Fig. 143. Left valve of *Modiomorpha obliqua* Cleland, 1911, Modiomorphoida: Modiomorphidae; PAL 80230; holotype; Berthelet Member; figured in Cleland, 1911, Pl 24, Fig 4 and Pohl, 1929, Pl 12, Fig 3; collected by EET; scale bar 5mm. Courtesy of Smithsonian Institution, Department of Paleobiology, National Museum of Natural History.

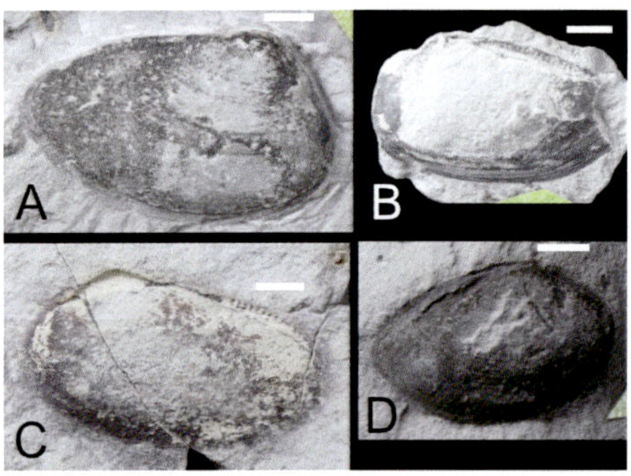

Fig. 144. *Palaeoneilo milwaukeensis* Cleland, 1911, Nuculoida: Malletiidae; Lindwurm Member; scale bar 5mm. **A:** Right valve; MPM 476; paratype; figured in Pohl, Pl 3, Fig 18; collected by CEM. **B:** Left valve; MPM 5012; North Avenue Intake Tunnel; paratype; figured in Pohl, 1929, Pl 3, Fig 22. **C:** Left valve; MPM P28162; collected by KCG. **D:** Right valve; MPM 2399; plesiotype; figured in Pohl, 1929, Pl 3, Fig 20.

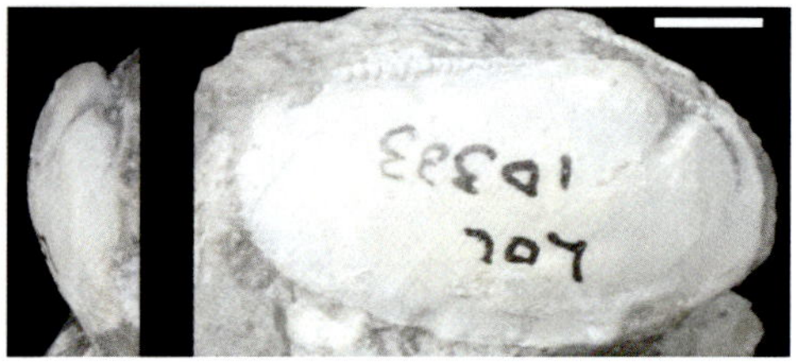

Fig. 145. Anterior and full view of silicified right valve of *Palaeoneilo milwaukeensis* Cleland, 1911, Nuculoida: Malletiidae; MPM 461; holotype; North Point Member; North Avenue Intake Tunnel; figured in Pohl, 1929, Pl 20, Fig 5; collected by CEM; scale bar is 5mm.

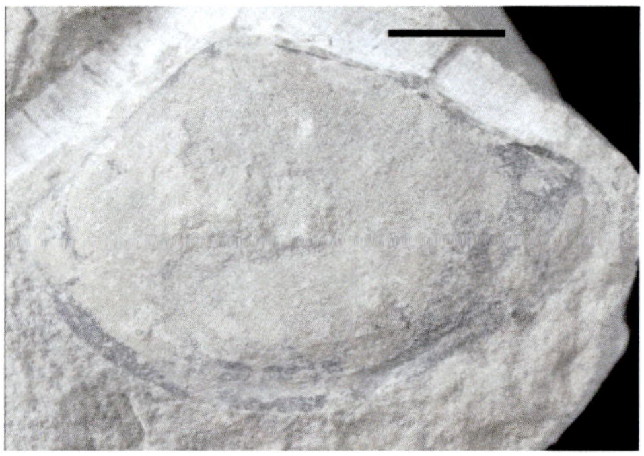

Fig. 146. Left valve of *Palaeoneilo devexa* Pohl, 1929, Nuculoida: Malletiidae; MPM 13576; paratype; Lindwurm Member; figured in Pohl, 1929, Pl 5, Fig 4; collected by GOR; scale bar 5mm.

Fig. 147. *Palaeoneilo corrugata* Pohl, 1929, Nuculoida: Malletiidae. **A:** Left valve; MPM 13579; paratype; Berthelet Member; figured in Pohl, 1929, Pl 4, Fig. 28; collected by GOR; scale bar 5mm. **B:** A silicified left valve; MPM 466; paratype; North Point Member; North Avenue Intake Tunnel; figured in Pohl, 1929, Pl 4, Fig 32; collected by CEM; scale bar 2mm. **C:** Right valve; MPM 468; paratype; figured in Pohl, 1929, Pl 4, Fig 29; collected by CEM; 2mm. **D:** Right valve; MPM 13546; syntype; figured in Pohl, 1929, Pl 4, Fig. 30; collected by GOR; scale bar 2mm. Bailey (1983) viewed *P. corrugata* as a synonym of *P. emarginata* (Conrad, 1841).

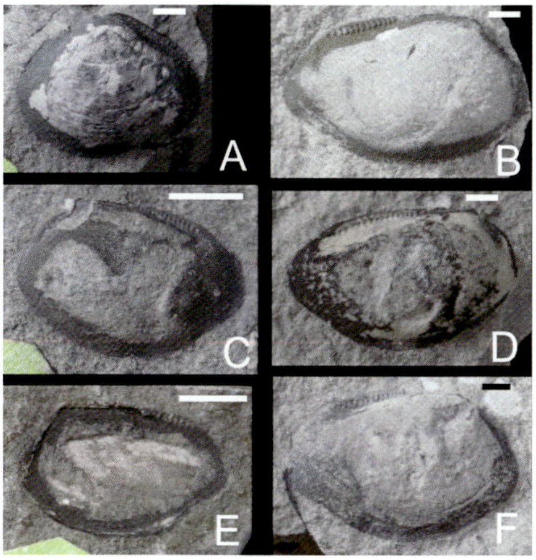

Fig. 148. *Palaeoneilo* sp. cf. *constricta* (Conrad, 1882), Nuculoida: Malletiidae; Berthelet Member. **A:** Left valve; MPM 463; figured in Cleland, 1911, Pl 20, Fig 22; collected by CEM; scale bar 2mm. **B:** Right valve; MPM 18580; figured in Pohl, 1929, Pl 4, Fig 8; collected by CEM; scale bar 2mm. **C:** Left valve; MPM 13559; figured in Pohl, Pl 4, Fig 4; collected by GOR; scale bar 5mm. **D:** Right valve; MPM 13555; figured in Pohl, 1929, Pl 4, Fig 2; collected by GOR; scale bar 2mm. **E:** Left valve; MPM 13532; figured in Pohl, 1929, Pl 20, Fig 3; collected by GOR; scale bar 5mm. **F:** Right valve; MPM 13556; figured in Pohl, 1929, Pl 5, Fig 15; collected by GOR; scale bar 2mm.

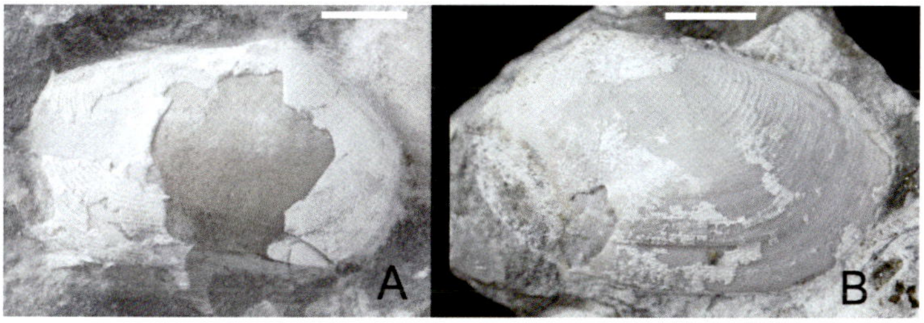

Fig. 149. Silicified valves of *Palaeoneilo pulchella* Pohl, 1929, Nuculoida: Malletiidae; syntypes; North Point Member; North Avenue Intake Tunnel; scale bars 5mm. **A:** Right valve; MPM 5159; figured in Pohl, 1929, Pl 4, Fig 47. **B:** Left valve; MPM 473; figured in Cleland, 1911, Pl 20, Fig 14 and Pohl, 1929, Pl 4, Fig 48; collected by CEM.

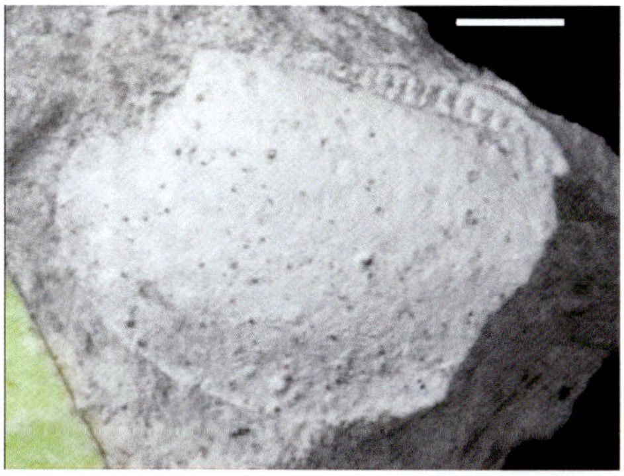

Fig. 150. Left valve of *Palaeoneilo dentata* Pohl, 1929, Nuculoida: Malletiidae; MPM 13619; syntype; North Point Member; figured in Pohl, 1929, Pl 4, Figs 19&24; collected by GOR; scale bar 2mm.

Fig. 151. Left valve, anterior, and dorsal views of *Palaeoneilo persinnuata* Pohl, 1929, Nuculoida: Malletiidae; MPM 13757; holotype; Lindwurm Member; figured in Pohl, 1929, Pl 4, Figs 35–37; collected by GOR; scale bar 5mm.

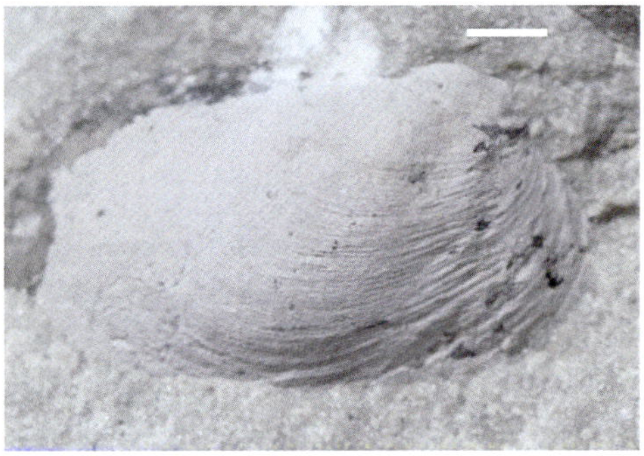

Fig. 152. Testiferous right valve of *Palaeoneilo* sp., Nuculoida: Malletiidae; MPM P28163; Berthelet Member; collected by KCG; scale bar 2mm.

Fig. 153. Anterior and right valve views of a possible *Nuculites laphami* Cleland, 1911, Nuculoida: Malletiidae; MPM P28164; Lindwurm Member; collected by KCG; scale bar 2mm.

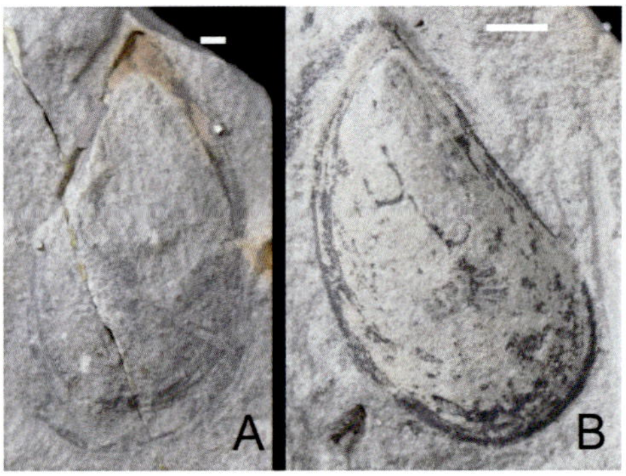

Fig. 154. *Mytilarca dentata* Pohl, 1929, Pterioida: Ambonychiidae; syntypes; Berthelet Member; collected by GOR; scale bars 2mm. **A:** Left valve; MPM 13539; figured in Pohl, Pl 5, Fig 21. **B:** Right valve; MPM 13574; figured in Pohl, Pl 5, Fig 20.

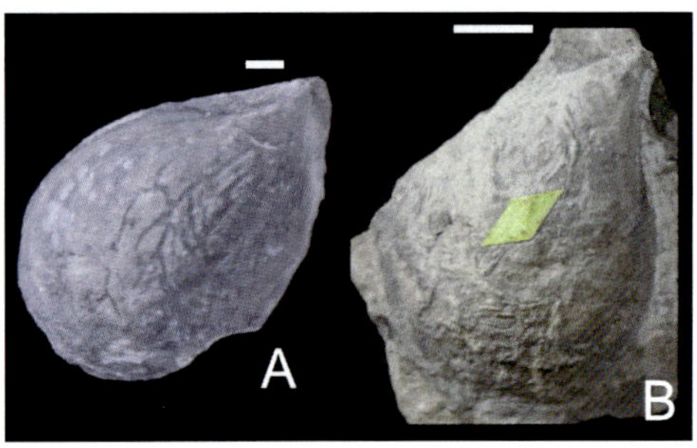

Fig. 155. *Mytilarca suberectus* (Pohl, 1929), Pterioida: Ambonychiidae Berthelet Member. **A:** Left valve; UWM-GM 4883; scale bar 5mm. **B:** Right valve; MPM 497; syntype; figured in Pohl, 1929, Pl 6, Fig 3; scale bar 1cm.

Fig. 156. Full and lateral views of left valve of *Mytilarca cingulosa* Pohl, 1929, Pterioida: Ambonychiidae; MPM 13580; syntype; Berthelet Member; figured in Pohl, 1929, Pl 5, Fig 15; collected by GOR; scale bar 5mm.

Fig. 157. *Mytilarca cingulosa* Pohl, 1929, Pterioida: Ambonychiidae; Berthelet Member; scale bar 5mm. **A:** Right valve; MPM 498; syntype; figured in Pohl, 1929, Pl 5, Fig 16; collected by CEM. **B:** Right valve; MPM 13537; syntype; figured in Pohl, 1929, Pl 5, Fig 19; collected by GOR. **C:** External mold of right and left valves; MPM 13538; syntype; figured in Pohl, 1929, Pl 5, Fig 18 (on left) and Fig 17 (on right); collected by GOR.

Fig. 158. Left valve of *Pterinopecten (Vertumnia) raaschi* (Pohl, 1929), Pterioida: Pterinopectenidae; MPM 13543; syntype; Berthelet Member; figured in Pohl, 1929, Pl 10, Fig 2; collected by GOR; scale bar 5mm.

Fig. 159. Right valve of *Pterinopecten (Vertumnia) intercalaris* (Pohl, 1929), Pterioida: Pterinopectenidae; MPM 13737; holotype; Lindwurm Member; figured in Pohl, 1929, Pl 9, Fig 5; collected by GOR; scale bar 5mm.

Fig. 160. Right valve of *Pterinopecten (Vertumnia) barretti* (Pohl, 1929), Pterioida: Pterinopectenidae; MPM 5156; holotype; Lindwurm Member; figured in Pohl, 1929, Pl 9, Fig 3; scale bar 5mm.

Fig. 162. Left valve of *Pterinopecten (Vertumnia) telleri* (Cleland, 1911), Pterioida: Pterinopectenidae; UWM-GM 5473; Berthelet Member; scale bar 5mm.

Fig. 161. Fragmentary right valve of *Pterinopecten (Vertumnia) edwardsi* (Pohl, 1929), Pterioida: Pterinopectenidae; MPM 13729; syntype; Lindwurm Member; figured in Pohl, 1929, Pl 9, Fig 2; collected by GOR; scale bar 5mm.

Fig. 164. Left valve of *Pterinopecten (Vertumnia)* sp., Pterioida: Pterinopectenidae; UWM-GM 5352; Berthelet Member; scale bar 5mm.

Fig. 163. Two images of the left valve of *Pterinopecten (Vertumnia) telleri* (Cleland, 1911), Pterioida: Pterinopectenidae. The image on the right has been highlighted with ammonium chloride sublimate to highlight the ornamentation and fine growth lines. PAL 80231; holotype; Berthelet Member; figured in Cleland, 1911, Pl 25, Fig 2 and Pohl, 1929, Pl 8, Fig 6; collected by EET; scale bar 5mm. Courtesy of Smithsonian Institution, Department of Paleobiology, National Museum of Natural History.

Fig. 165. External mold of left valve of *Ptychopteria (Actinopteria) rhombolineare* (Pohl, 1929), Pterioida: Pterinopectenidae; MPM 455; syntype; Lindwurm Member; figured in Pohl, 1929, Pl 7, Fig 13; collected by CEM; scale bar 5mm.

Fig. 166. Dorsal and full left valve views of *Ptychopteria (Actinopteria) rhombolineare* (Pohl, 1929), Pterioida: Pterineidae; MPM 456; syntype; Lindwurm Member; figured in Pohl, 1929, Pl 7, Fig 12; collected by CEM; scale bar 5mm.

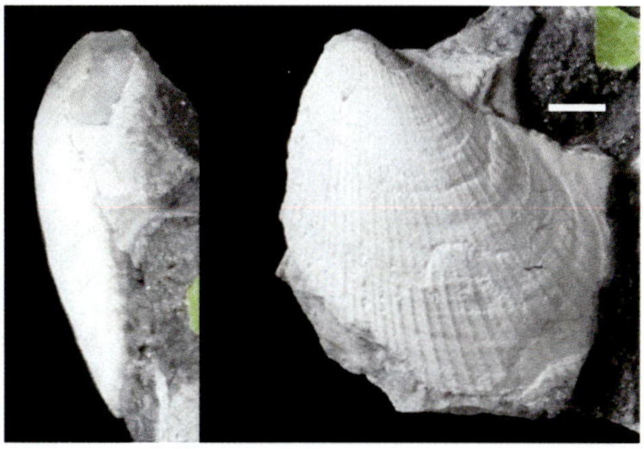

Fig. 167. Dorsal and full left valve views of *Ptychopteria (Actinopteria) singularis* (Pohl, 1929), Pterioida: Pterineidae; MPM 2301; holotype; North Point Member; figured in Pohl, 1929, Pl 7, Fig 16; scale bar 2mm.

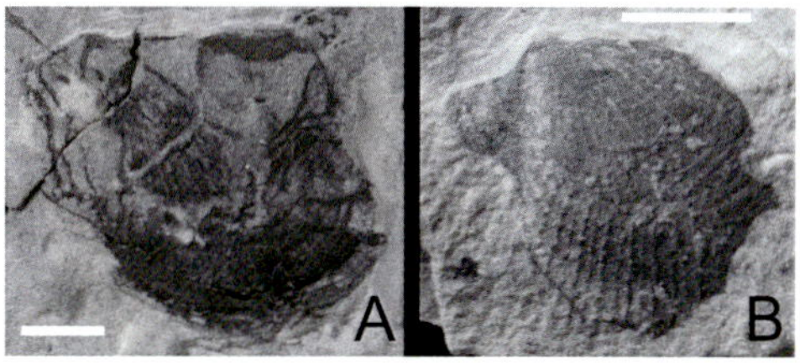

Fig. 168. *Ptychopteria (Actinopteria) pauciradiata* (Pohl, 1929), Pterioida: Pterineidae; left valves; Berthelet Member; scale bar 5mm. **A:** MPM 2302; holotype; figured in Pohl, 1929, Pl 7, Fig 14. **B:** MPM 5149; paratype; figured in Pohl, 1929, Pl 7, Fig 15.

Fig. 169. Left valve of *Leptodesma (Leiopteria) acutilaris* (Pohl, 1929), Pterioida: Pterineidae; MPM 13740; syntype; Lindwurm Member; figured in Pohl, 1929, Pl 5, Fig 9; collected by GOR; scale bar 5mm.

Fig. 170. Right valve of *Aviculopecten duplicostatus* Pohl, 1929, Pterioida: Aviculopectenidae; MPM 13529; holotype; Berthelet Member; figured in Pohl, 1929, Pl 8, Fig 4; collected by GOR; scale bar 5mm.

Fig. 171. Right valve of *Eoschizodus dubius* (Pohl, 1929), Trigonioida: Myophoriidae; MPM 13572; holotype; Berthelet Member; figured in Cleland, 1911, Pl 10, Fig 4; collected by GOR; scale bar 5mm.

Fig. 173. Left valve of *Paracyclas paradoxica* Pohl, 1929, Heterodonta: Mactromyidae; MPM 13622; holotype; North Point Member; figured in Pohl, 1929, Pl 10, Fig 5; collected by GOR; scale bar 5mm.

Fig. 175. Anterior and full view of left valve of *Cimitaria obtusiloba* (Cleland, 1911), Pholadomyoida: Grammysiidae; MPM 10224; Berthelet Member; figured in Pohl, 1929, Pl 2, Fig 4; scale bar 5mm.

Fig. 172. Right valve of *Aviculopecten* sp., Pterioida: Aviculopectenidae; MPM 13730; Lindwurm Member; figured in Pohl, 1929, Pl 8, Fig 5; collected by GOR; scale bar 5mm.

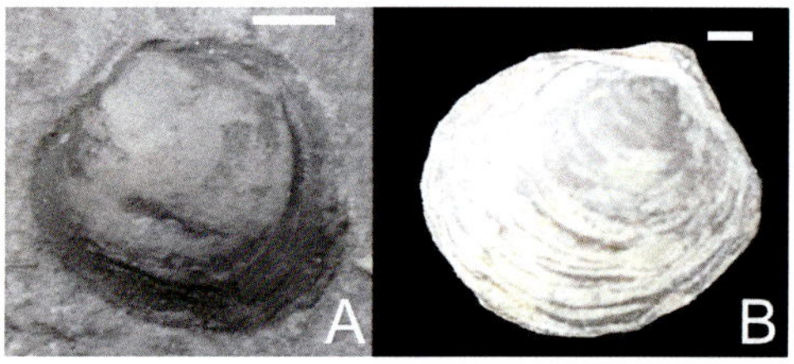

Fig. 174. *Paracyclas* sp., Heterodonta: Mactromyidae; collected by CEM; scale bar 2mm. **A:** Left valve; MPM 500; figured in Cleland, 1911, Pl 25, Fig 8 and Pohl, 1929, Pl 14, Fig 24; Berthelet Member. **B:** Right valve; MPM 0493; figured in Cleland, 1911, Pl 25, Figs 6&7 and Pohl, 1929, Pl 14, Fig 23; Lindwurm Member.

Fig. 176. Right valve of *Cimitaria obtusiloba* (Cleland, 1911), Pholadomyoida: Grammysiidae; MPM 10225; holotype; Berthelet Member; figured in Cleland, 1911, Pl 25, Fig 11; scale bar 5mm.

Fig. 179. Left valve and dorsal views of *Grammysia* sp., Pholadomyoida: Grammysiidae; UWM-GM 4693; Berthelet Member; scale bar 5mm.

Fig. 177. Three views of *Grammysia nodocostata* Hall, 1885, Pholadomyoida: Grammysiidae; MPM 478; Berthelet Member; figured in Cleland, 1911, Pl 23, Figs 1&2; collected by Milwaukee Cement Company; scale bar 1cm.

Fig. 178. Anterior and full views of fragmentary left valve of *Grammysia marginata* Pohl, 1929, Pholadomyoida: Grammysiidae; MPM 477; syntype; Berthelet Member; figured in Cleland, 1911, Pl 23, Fig 3 and Pohl, 1929, Pl 1, Fig 6; collected by CEM; scale bar 5mm.

Fig. 180. Anterior view of *Grammysia ulrichi* Cleland, 1911, Pholadomyoida: Grammysiidae; MPM 20648; Lindwurm Member; collected by BB; scale bar in mm.

Fig. 181. Right valve and anterior views of *Grammysia ulrichi* Pohl, 1929, Pholadomyoida: Grammysiidae; UWM-GM 1424; Berthelet Member; scale bar 1cm.

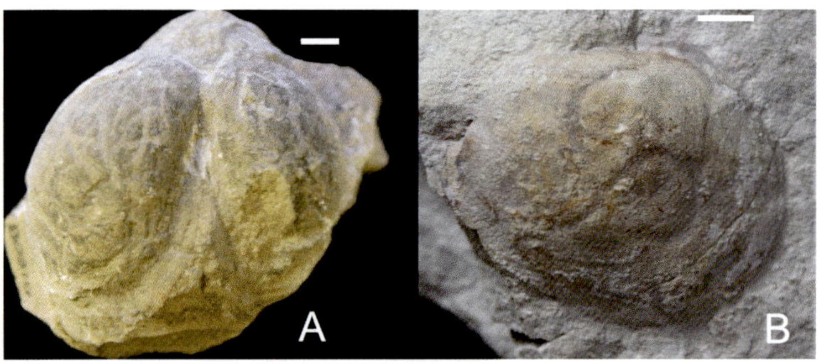

Fig. 182. *Edmondia fragilis* Cleland, 1911, Pholadomyoida: Edmondiidae; Berthelet Member; scale bars 5mm. **A:** PAL 80310; syntype; figured in Cleland, 1911, Pl 21, Fig 1; collected by EET; courtesy of Smithsonian Institution, Department of Paleobiology, National Museum of Natural History. **B:** Left valve; MPM P28165; collected by KCG.

Cephalopoda (Cambrian–Holocene)

The *Treatise on Invertebrate Paleontology* describes cephalopods as "agile, bilaterally symmetrical, predacious [molluscan] carnivores that achieved a structural complexity and metabolic efficiency without peer among unsegmented invertebrates" (Sweet, 1964). Among the typical specialized features of living cephalopod species are a pair of complex eyes, a beak, a set of eight or more elongated, flexible appendages, and a hyponome (or funnel) for propulsion through the water. Cephalopods have either a single calcareous external or internal shell or may lack a shell entirely.

Most of today's cephalopods belong to the subclass Coleoidea, which has eight or ten sucker or hook bearing appendages and a single pair of gills. They include octopuses, argonauts, squids, cuttlefish, and similar forms. Except for the argonauts, all of those coleoids plus the extinct belemnoids have a shell that was either internal or lacking. None of these groups have been found in rocks deposited prior to the Carboniferous Period.

Instead of coleoids, most of the cephalopods from the Milwaukee Formation consist of the sturdy external shells of species from the Actinoceratoidea and Nautiloidea, while two shells that appear to have been from the Ammonoidea were also collected, as shown at the end of this chapter. Cephalopods from those three subclasses had shells that were conical, coiled or of various other shapes, and were divided into chambers, which the animal added as it grew by secreting a calcareous septum between the chambers. A tube (siphuncle) running through the chambers was used to regulate buoyancy by pumping ions into or out of the siphuncle which allows watery fluids or nitrogenous gas to be soaked up or given off to chambers, as the living *Nautilus* still does. Sutures mark the position of the septa on the exterior surfaces of the shells (Fig. 183). The sutures of ammonoids tend to be elaborate in outline. Taxonomists use their designs to distinguish ammonoids from other cephalopods and from each other.

The most common cephalopods from the Milwaukee Formation are the nautiloids here referred to as *Acleistoceras?* but probably representing several genera. This genus is currently divided into several species differentiated largely by the outline of their shells and the length of their living chambers. Identification based on these criteria is difficult due to the variable distortion of most specimens and limited knowledge of their living chambers. Fossils of these species, especially *A.? wisconsinense* (Cleland, 1907), can take on tremendous variation in form, depending on the orientation of the shell on the substrate during burial and the degree of vertical compression (Figs. 184–188). Cleland recognized this phenomenon, but observed that some of these fossils were nearly circular in cross-sectional outline. He thus built plaster reconstructions of the conchs of four species, showing their suggested shapes prior to distortion (Cleland, 1907, 1911). Those models are shown here (Fig. 189). Cleland was quick to point out that his interpretation of the length of the living chamber, and the shape of the shell's posterior extremity and of the hyponomic sinus were hypothetical for most of the taxa. Today the collections of these cephalopods in the Milwaukee Public Museum and Greene Museum are extensive and should therefore permit a fruitful modern analysis. It is also noteworthy that fossils of *A.? wisconsinense* suggest an overall valve length of up to 36cm and that the cover image showing this species swimming apex-up is purely conjectural. Cleland (1911) suggested that it may indeed never have left the substrate.

The cephalopod fauna of the Milwaukee Formation is one of the most diverse known from the Middle Devonian rocks of the Midwest and is represented by an exceptionally large number of specimens in museums. Nearly all of these specimens were collected from the top six feet of Raasch's Berthelet Member, which is equivalent to Zone B of Cleland. This interval has informaly been called the cephalopod beds. Cephalopods are less common in the Lindwurm Member.

Most of the cephalopod species from the Milwaukee Formation are illustrated here (Figs. 190–205). Cleland (1911) included *Orthoceras* sp. and *Orthoceras clarkei* Cleland, 1911 as well.

Fig. 183. Shell of the nautiloid cephalopod *Gyronaedyceras eryx* (Hall, 1879) with selected external morphological features and labeled; Oncocerida: Brevicoceratidae; MPM 437; Berthelet Member; figured in Cleland, 1907, Fig 14 and Cleland, 1911, Pl 33; collected by CEM; scale bar 2cm.

Fig. 184. *Acleistoceras? wisconsinense* (Cleland, 1907), Oncocerida: Acleistoceratidae; MPM 407; syntype; Berthelet Member; figured in Cleland, 1907, Fig 3 and Cleland, 1911, Pl 31; collected by CEM; scale bar 2cm.

Fig. 185. *Acleistoceras? wisconsinense* (Cleland, 1907), Oncocerida: Acleistoceratidae; the least-distorted nearly complete example of this species. Note the constricted band at the back of the living chamber, characteristic of this species. MPM 411; syntype; Berthelet Member; figured in Cleland, 1907, Fig 1 and Cleland, 1911, Pl 29; collected by CEM; scale bar 2cm.

Fig. 186. *Acleistoceras? wisconsinense* (Cleland, 1907), Oncocerida: Acleistoceratidae; another view of the specimen from the previous page, showing the degree of flattening; MPM 411; scale bar 2cm.

Fig. 187. An extreme example of a symmetrically flattened *Acleistoceras? wisconsinense* valve, due to a vertical orientation of its long axis during burial; note the concentricity of the sutures; UWM-GM 1117; Berthelet Member; scale bar 1cm.

Fig. 188. *Acleistoceras? wisconsinense* (Cleland, 1907) Oncocerida: Acleistoceratidae; flattened, with apex near top-center of image; MPM 418; Berthelet Member; collected by CEM; scale bar 1cm.

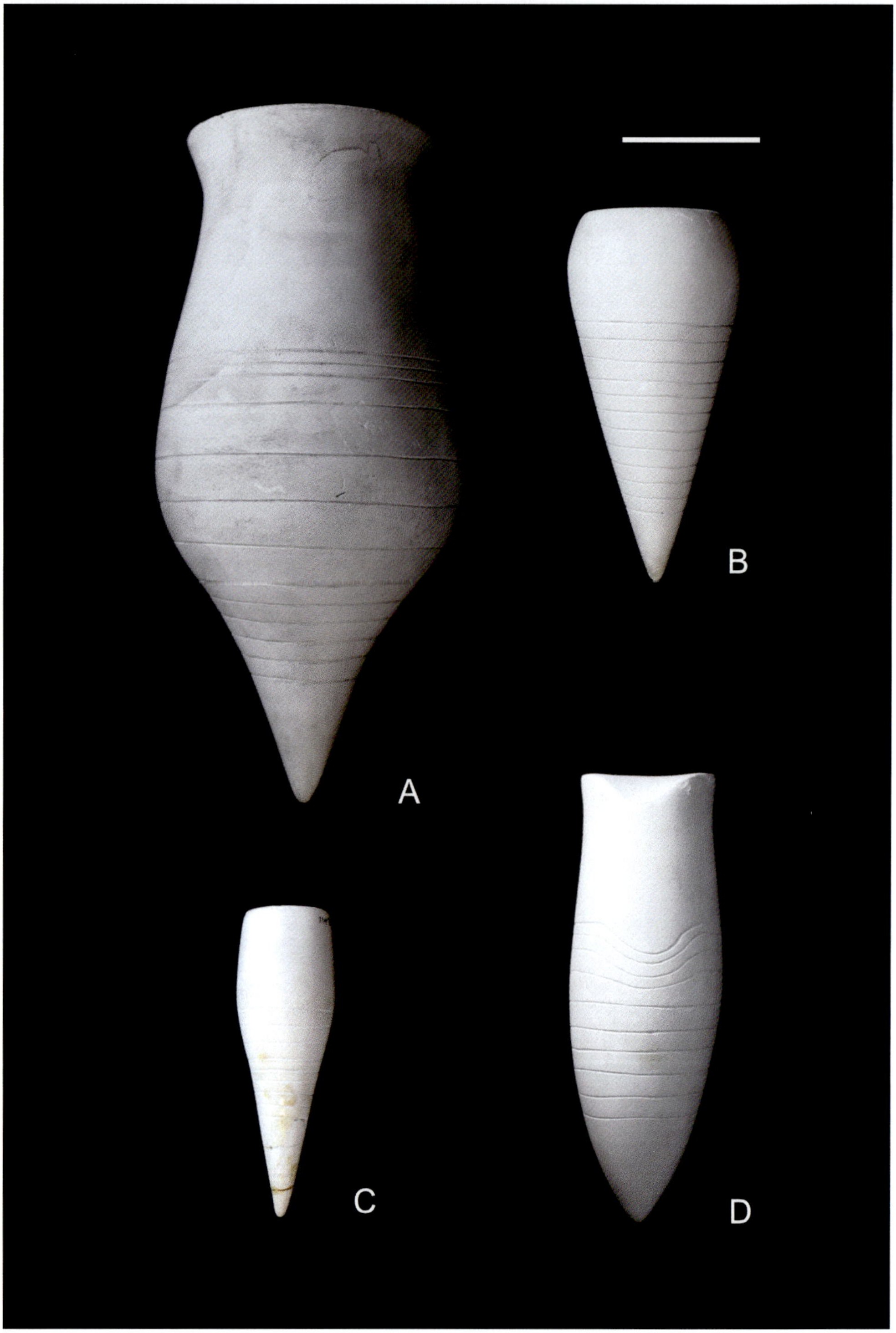

Fig. 189. Cleland's plaster models of certain cephalopod species from the Milwaukee Formation showing his interpretation of the original shapes; each is circular in cross-sectional outline; lateral views; scale bar 5cm. **A:** *Acleistoceras? wisconsinense* (Cleland, 1907); MPM 11462; figured in Cleland, 1907, Fig 2, and Cleland, 1911, Pl 30. **B:** *Acleistoceras? whitfieldi* (Cleland, 1907); MPM 11479; figured in Cleland, 1907, Fig. 9. **C:** *Acleistoceras? fusiforme* (Whitfield, 1882); MPM 11475; figured in Cleland, 1907, Fig 12, and Cleland, 1911, Pl 34, Fig 2. **D:** *Acleistoceras? calvini* (Cleland, 1907); MPM 11463; figured in Cleland, 1907, Fig 7, and Cleland, 1911, Pl 34, Fig 1.

Fig. 190. *Acleistoceras? fusiforme* (Whitfield *in* Chamberlin, 1882), Oncocerida: Acleistoceratidae; PAL 137047; holotype; Berthelet Member; figured in Whitfield, 1882, Pl 26, Fig 16; collected in Whitefish Bay; scale bar 5mm. Courtesy of Smithsonian Institution, Department of Paleobiology, National Museum of Natural History. The catalog entry indicates EET as collector; however, the label on the specimen suggests it may have been obtained from I.A. Lapham.

Fig. 191. *Acleistoceras? fusiforme* (Whitfield *in* Chamberlin, 1882), Oncocerida: Acleistoceratidae; two views showing the typical degree of flattening that most cephalopods in the Milwaukee Formation underwent; MPM 432; Berthelet Member; figured in Cleland, 1907, Fig 11 and Cleland, 1911, Pl 41, Fig 1; collected by CEM; scale bar 1cm.

Fig. 192. *Acleistoceras? breviposticum* (Whitfield *in* Chamberlin, 1882), Oncocerida: Acleistoceratidae; PAL 137048; holotype; Berthelet Member; figured in Whitfield, 1882, Pl 26, Fig 15; collected at Whitefish Bay; scale bar 5mm. Courtesy of Smithsonian Institution, Department of Paleobiology, National Museum of Natural History. The catalog entry indicates EET as collector; however, the specimen was probably obtained from T.C. Chamberlin or I.A. Lapham.

Fig. 194. *Acleistoceras? calvini* (Cleland, 1907), Oncocerida: Acleistoceratidae; MPM 11465; Berthelet Member; collected by GOR; scale bar 1cm.

Fig. 193. *Acleistoceras? calvini* (Cleland, 1907), Oncocerida: Acleistoceratidae; MPM 3156; Berthelet Member; scale bar 1cm.

Fig. 195. *Acleistoceras? calvini* (Cleland, 1907), Oncocerida: Acleistoceratidae; MPM 4698; Berthelet Member; scale bar 1cm.

Fig. 197. *Acleistoceras? whitfieldi* (Cleland, 1907), Oncocerida: Acleistoceratidae; MPM 3186; Berthelet Member; scale bar 1cm.

Fig. 196. *Acleistoceras? whitfieldi* (Cleland, 1907), Oncocerida: Acleistoceratidae; UWM-GM 9475; Berthelet Member; scale bar 1cm.

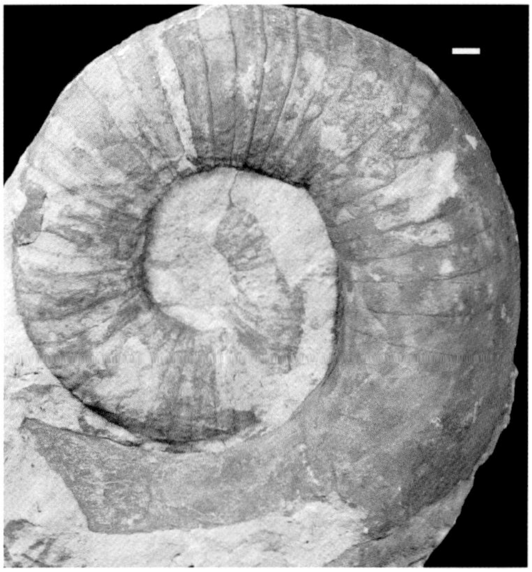

Fig. 198. *Gyronaedyceras eryx* (Hall, 1879), Oncocerida: Brevicoceratidae; UWM-GM 25773; Berthelet Member; scale bar 1cm. This species was originally placed in *Gyroceras* followed by *Rhadinoceras*, and finally was included in *Gyronaedyceras* partially because of its marginal siphuncle (Flower & Gordon, 1972).

Fig. 199. *Gyronaedyceras eryx* (Hall, 1879), Oncocerida: Brevicoceratidae; UWM-GM 25782; Berthelet Member; scale bar 1cm.

Fig. 200. *"Orthoceras"* sp.; UWM-GM 853; Berthelet Member; scale bar 1cm.

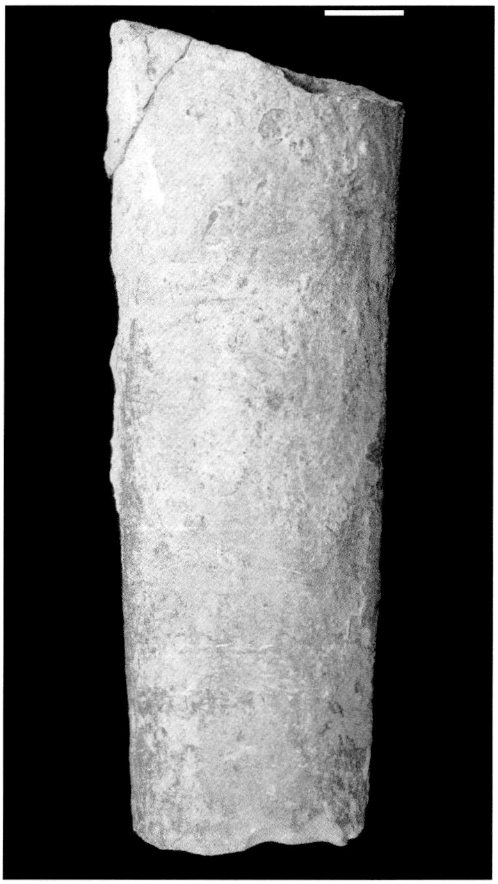

Fig. 201. *Ormoceras* cf. *O. bebryx* (Hall, 1879), Actinocerida: Ormoceratidae; MPM 490; Berthelet Member; figured in Cleland, 1911, Pl 43, Figs 1&2; collected by Milwaukee Cement Company; scale bar 2cm.

Fig. 202. *"Orthoceras"* sp; posterior end broken off cleanly at septum; MPM 491; Berthelet Member; collected by CEM; scale bar 1cm.

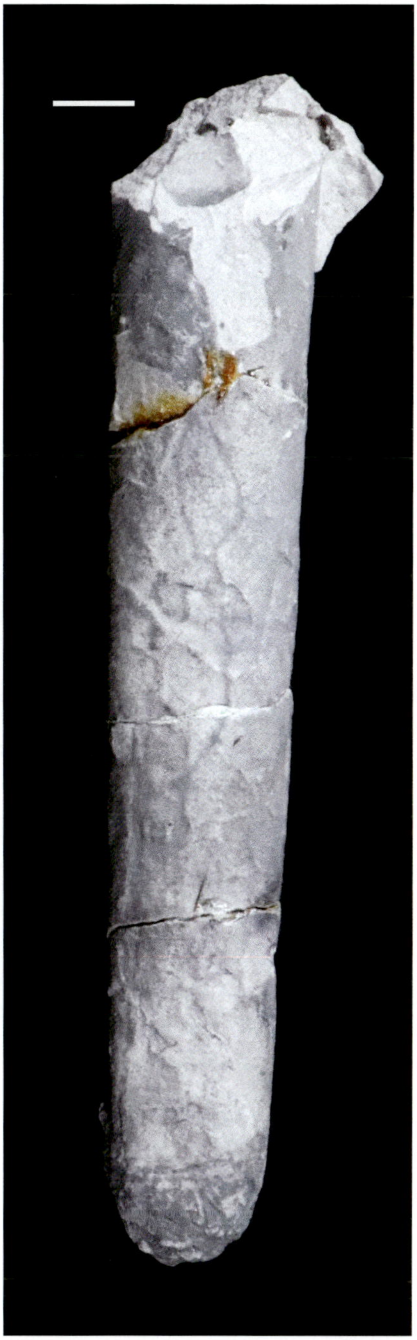

Fig. 203. An *"Orthoceras"* valve showing very little narrowing posteriorly in comparison to other *"Orthoceras"* specimens from the Milwaukee Formation; UWM-GM 48211; Berthelet Member; scale bar 1cm.

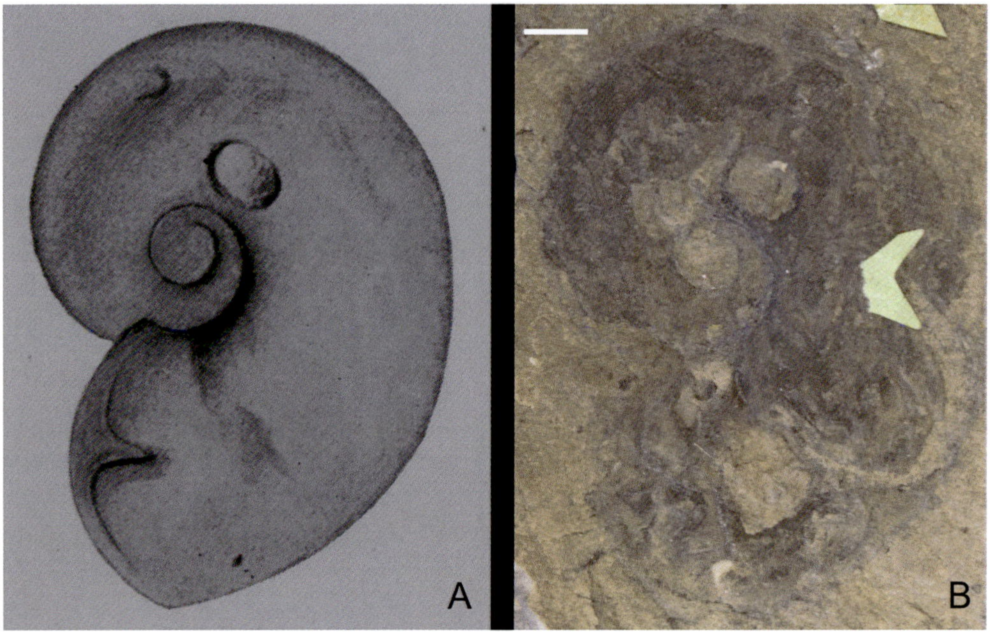

Fig. 204. Possible ammonoid cephalopod; Berthelet Member; scale bar 5mm. **A:** Photograph of drawing from Cleland, 1911, Pl 42, Fig 4. Cleland referred to this specimen as "*Goniatites* sp. undet." Courtesy of Wisconsin Geological and Natural History Survey. **B:** The specimen (PAL78336) on which Cleland based the drawing. Were this poorly preserved fossil and its imaginative restoration the only evidence of ammonoids in Wisconsin, the presence of that group may not have been credible. Courtesy of Smithsonian Institution, Department of Paleobiology, National Museum of Natural History.

Fig. 205. Ventral and lateral views of what appears to be an ammonoid cephalopod. Greene's label and the catalog entry for this specimen read, "Prof Hall says *Goniatites*...unlike anything I have obtained previously." UWM-GM 5449; Berthelet Member; scale bar 5mm.

Chapter 15 — Annelid Worms
(Annelida)

The Annelida (Cambrian–Holocene) is the phylum of protostome animals consisting of aquatic or terrestrial, vermiform, bilaterally symmetrical, coelomate metazoans that have a segmented external cuticle that is never molted. It includes earthworms, leeches, marine worms and many other worm-like animals. Most are infaunal filter feeders or benthic predators.

The fossil evidence of annelids in the Milwaukee Formation includes trace fossils (Chapter 23) and **scolecodonts**, which are the chitinous jaws of polychaete worms.

Polychaeta (Cambrian–Holocene)

The Polychaeta is a diverse class of mostly marine annelids that typically bear unjointed appendages that develop as extensions of the animal's body segments. As with foraminifers and other microfossils, scolecodonts are easily missed in collections, due to their small size. Consequently, relatively few scolecodonts in existing museum collections that house Milwaukee Formation fossils are labeled or catalogued. The only species thus far identified from this formation is *Arabellites crescentum* Cleland, 1911, represented only by scolecodonts (Fig. 206). Other taxa were sure to exist. Recovering some of them is likely with further examination of the existing collections.

Fig. 206. *Arabellites crescentum* Cleland, 1911, Errantida: Leodicidae; PAL 85990; holotype; Lindwurm Member; figured in Cleland, 1911, Pl 4, Fig 1; collected by EET; scale bar 1mm. Courtesy of Smithsonian Institution, Department of Paleobiology, National Museum of Natural History.

Chapter 16 — Arthropods
(Arthropoda)

The Arthropoda (Precambrian?, Cambrian–Holocene) is the phylum of protostome metazoans that has a chitinous exoskeleton and typically segmented body with jointed legs. It includes the chelicerates, trilobites, myriapods, crustaceans, insects, and several other living and extinct aquatic and terrestrial groups. The Arthropoda is the most diverse phylum of animals alive today as it probably was through much of Phanerozoic history. The two subphyla that are represented in the Milwaukee Formation are the Trilobita and Crustacea.

Trilobita (Cambrian–Permian)
Trilobites make up a common and extremely diverse subphylum of marine Paleozoic arthropods characterized by a calcite-impregnated exoskeleton divided longitudinally into three lobes (Whittington et al., 1997) (Fig. 207). Like other arthropods, their exoskeletons were segmented and were shed as the animals grew. Most of the trilobite fossils in the Milwaukee Formation are disarticulated pieces of the exoskeleton—mainly cephala, pygidia, and thoracic segments.

The trilobite fauna from the Milwaukee Formation consists only of benthic species. The abundance of predatory fish and cephalopods swimming in the Milwaukee waters was undoubtedly a powerful deterrent against nectonic trilobites and probably helps account for the relatively few fossils of the trilobite species that were present. The lack of spinose species is consistent with relatively high energy environments, such as that under which the Milwaukee Formation was deposited. Those environments favored trilobites that were compact, enrolled tightly, and were good at fitting beneath bryozoan mounds and other high spots during stormy periods (R. Fortey, personal communication, 19-Jul-2018). Phacopids and proetids fit those criteria well and are in fact the only trilobite fossils present in the Milwaukee Formation biota.

The most common trilobite in the Lindwurm and Berthelet Members is *Eldredgeops*? *norwoodensis* (Stumm, 1953). Articulated specimens are uncommon but when present are usually preserved in varying degrees of enrollment (Figs. 208–212). A specimen from the Milwaukee Formation is a paratype of this species (Stumm, 1953). *E.*? *norwoodensis* specimens from this formation were used to help support the theory of punctuated equilibria (Eldredge, 1972; Eldredege & Gould, 1972; Brown, 1987). In that theory, Eldredge & Gould argued that evolution is not always slow and steady, but can proceed in long periods of stasis followed by sudden bursts of morphological change. They used variation in eye morphology (number of vertical dorso-ventral files of lenses) of *Phacops*, which was then considered to include the species now generally included in *Eldredgeops*?, as a primary example.

It is noted that the genus *Eldredgeops* appears to be a better fit for *norwoodensis* and related species than is *Phacops* (Struve, 1990; Burton & Eldredge, 1974; Kloc, 2014), although formal phylogenetic analysis is needed to verify this assignment (Niles Eldredge, personal communication, 31-Aug-2018; B.S. Lieberman, personal communication, 01-Sep-2018).

Lacking glabellar tubercles and having a flat border on its pygidium, the proetid trilobites found in the Milwaukee Formation are easy to distinguish from *Eldredgeops*? (Figs. 213 & 214). The free cheeks of proetids are generally found cleanly disarticulated from the cranidium (Fig. 215), unlike the free cheeks of *Eldredgeops*?, which generally remain fused to the cranidium during molting.

Cleland's (1911) discussion of the Wisconsin Devonian proetid trilobites included only *Proetus rowi* Green, 1838 (a species currently assigned to *Pseudodechenella*) and possible occurrences of *Proetus crassimarginatus* Hall, 1843 (a species currently assigned to *Crassiproetus*). While some specimens in the collections of the Milwaukee Public Museum and Greene Geological Museum do appear to belong to *Crassiproetus* (Fig. 214) and *Pseudodechenella*? (Fig. 215 D), and others are identified as *Dechenella*?

(Fig. 215 A&B), further preparation and analysis of the material in the various collections is needed to sort out what proetid taxa are truly present. *C. crassimarginatus* is found in the Little Cedar Formation of Iowa; however, the length/width ratio of its pygidium appears to be larger than that of the Wisconsin form.

The proetids of the Milwaukee Formation have only been found in the Lindwurm Member, yet *Eldredgeops*? Is found primarily as cephala or pygidia in both the upper Berthelet and Lower Lindwurm. The percentage of outstretched *Eldgedgeops*? exoskeletons (as opposed to enrolled ones) appears to be greater in the Berthelet Member than in the Lindwurm (Fig. 210 E). The authors know of no trilobite taxa from the Milwaukee Formation other than *Eldredgeops*? *norwoodensis* and proetids.

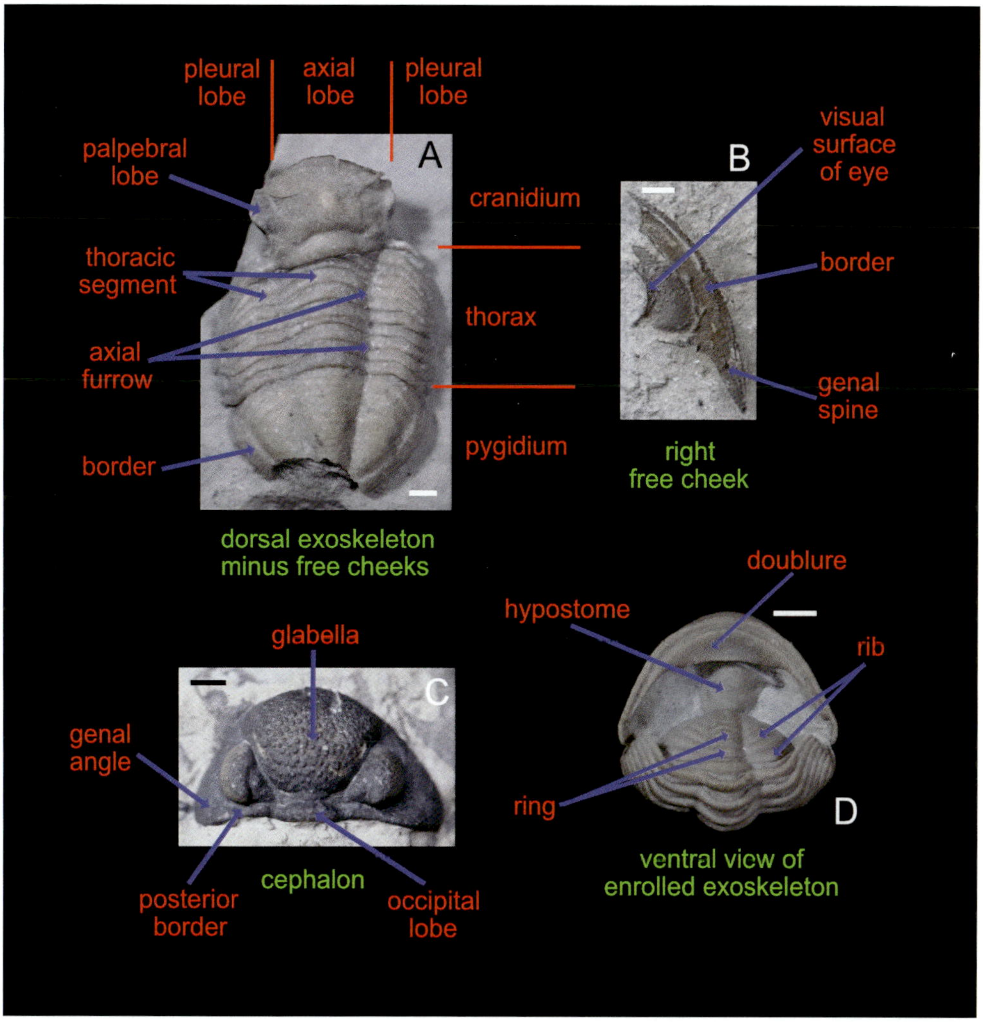

Fig. 207. Trilobite exoskeletal elements with selected external morphological features labeled; Lindwurm Member, except B, which is probably from the North Point Member. **A:** Proetid; probably *Crassiproetus* sp.; MPM P365901; scale bar 2mm. **B:** Proetid; MPM 5120; scale bar 2mm. **C:** *Eldredgeops? norwoodensis* (Stumm, 1953); scale bar 2mm. **D:** *E.? norwoodensis*; WESM-KCG 1126; scale bar 5mm.

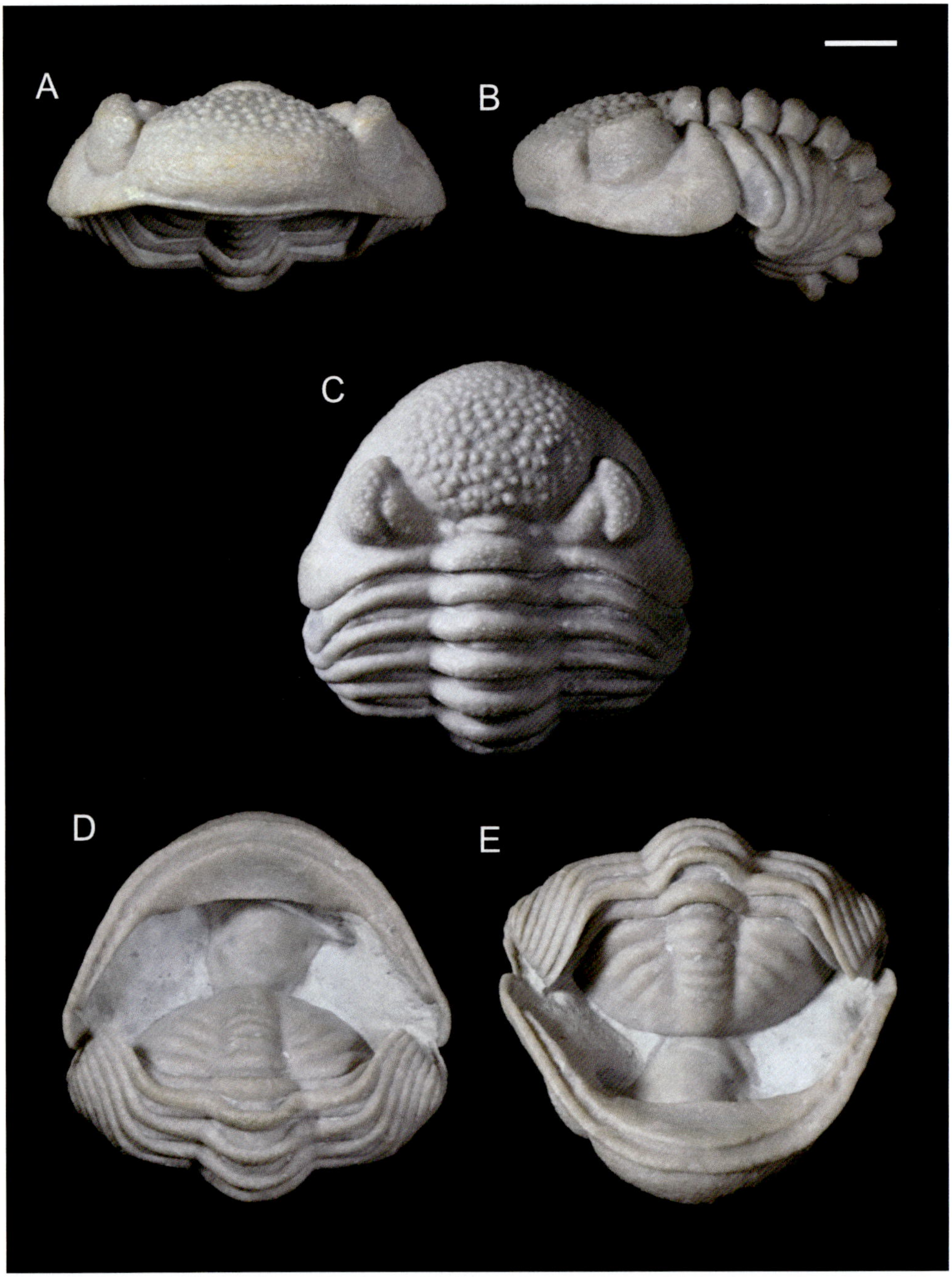

Fig. 208. Five views of a large, well preserved, complete, partially enrolled exoskeleton of *Eldredgeops? norwoodensis* (Stumm, 1953), Phacopida: Phacopidae; WESM-KCG 1126; Lindwurm Member; collected by KCG; scale bar 5mm. **A:** Anterior view. **B:** Lateral view. **C:** Dorsal view of cephalon and the visible portion of thorax. **D:** Ventral view of cephalon showing the detached hypostome partially obscured by the cephalic doublure and pygidium. **E:** View showing pygidium with partially-detached eleventh thoracic segment. Prepared by David Comfort.

Fig. 209. Dorsal, anterior and lateral views of a cephalon of *Eldredgeops? norwoodensis* (Stumm, 1953), Phacopida: Phacopidae; R. Kneusel collection, uncatalogued; Lindwurm Member; collected by RK; scale bar 2mm; prepared by David Comfort.

Fig. 210. *Eldredgeops? norwoodensis* (Stumm, 1953), Phacopida: Phacopidae; scale bars 5mm. **A:** Thorax; MPM P28166; Lindwurm Member; collected by KCG. **B:** Enrolled exoskeleton; MPM 25834; Lindwurm Member; collected by DGM. **C:** Thorax and pygidium; MPM 25843; Lindwurm Member; collected by DGM. **D:** Anterior view of cephalon of enrolled exoskeleton; MPM 25929; Lindwurm Member; collected by RO. **E:** Internal mold of exoskeleton; UWM-GM 3954; Berthelet Member. **F:** Internal mold of cephalon; MPM P26724; Berthelet Member.

Fig. 211. Two views of a partially enrolled exoskeleton of *Eldredgeops? norwoodensis* (Stumm, 1953), Phacopida: Phacopidae; MPM P28167; Lindwurm Member; collected by KCG; scale bar 5mm.

Fig. 212. Partially enrolled exoskeleton of *Eldredgeops? norwoodensis* (Stumm, 1953), Phacopida: Phacopidae; MPM 16675; Lindwurm Member; collected by GOR; scale bar 5mm.

Fig. 213. Side-by-side comparison of the pygidia of phacopid (top) and proetid (bottom) trilobites. WESM-KCC 1048; Lindwurm Member; collected by JSC; scale bar 5mm.

Fig. 214. Lateral and dorsal views of *Crassiproetus* sp. lacking librigenae and with a displaced, crushed cranidium; Proetida: Proetidae. Even in this condition, this is one of the most complete proetid specimens known from the Milwaukee Formation. MPM P365901; Lindwurm Member; collected by LAG; photograph courtesy of Joshua Gass; scale bar 2mm; prepared by David Comfort.

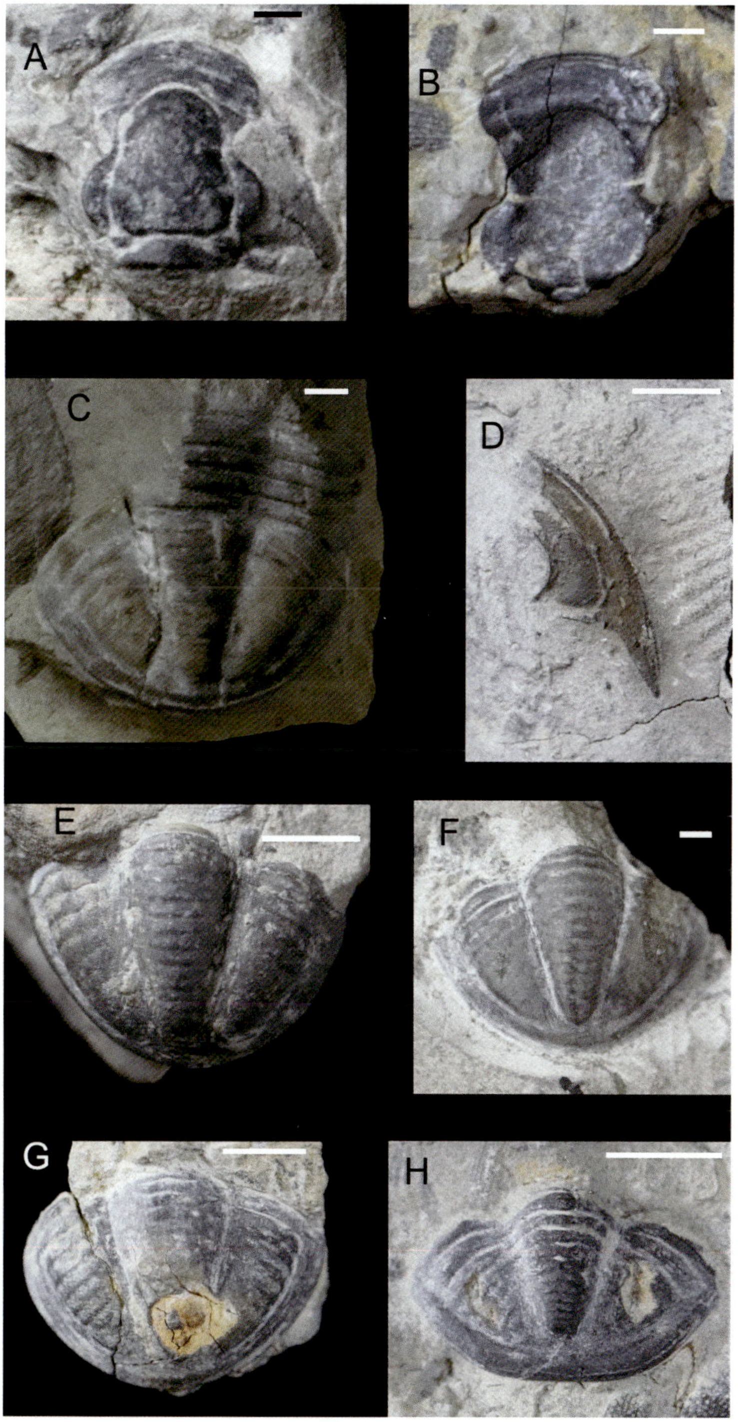

Fig. 215. Eight pieces of exoskeletons of proetids, Proetida: Proetidae; Lindwurm Member, except D, which is probably from the North Point Member. **A:** Partial cranidium of *Dechenella?* sp.; MPM 16716; scale bar 2mm. **B:** Interior of partial cranidium of *Dechenella?* sp; MPM 16717; collected by GOR; scale bar 5mm. **C:** Pygidium with part of thorax attached; MPM 16705; collected by GOR; scale bar 2mm. **D:** Right free cheek of *Pseudodechenella?* sp.; MPM 5120; North Avenue Intake Tunnel; scale bar 5mm. **E:** Pygidium; MPM P28169; collected by KCG; scale bar 5mm. **F:** Pygidium; MPM 16700; collected by GOR; scale bar 2mm. **G:** Pygidium of *Crassiproetus?* sp.; MPM 10695; collected by GOR; scale bar 5mm. **H:** Unusually wide pygidium, but with the flattened border typical of proetids; MPM 16711; collected by GOR; scale bar 5mm.

Phyllocarida (Cambrian–Holocene)

The Phyllocarida are malacostracan crustaceans whose head and thoracic regions are encased in a bivalved chitinous carapace (Rolfe, 1969). The Teller collection has fossils of the phyllocarids *Paraechinocaris punctata* (Hall, 1863) (Fig. 216) and *Tropidocaris* sp. (Cleland, 1911, Pl. 44, Fig. 9) from the Milwaukee Formation. No other phyllocarid taxa have thus far been identified from this formation.

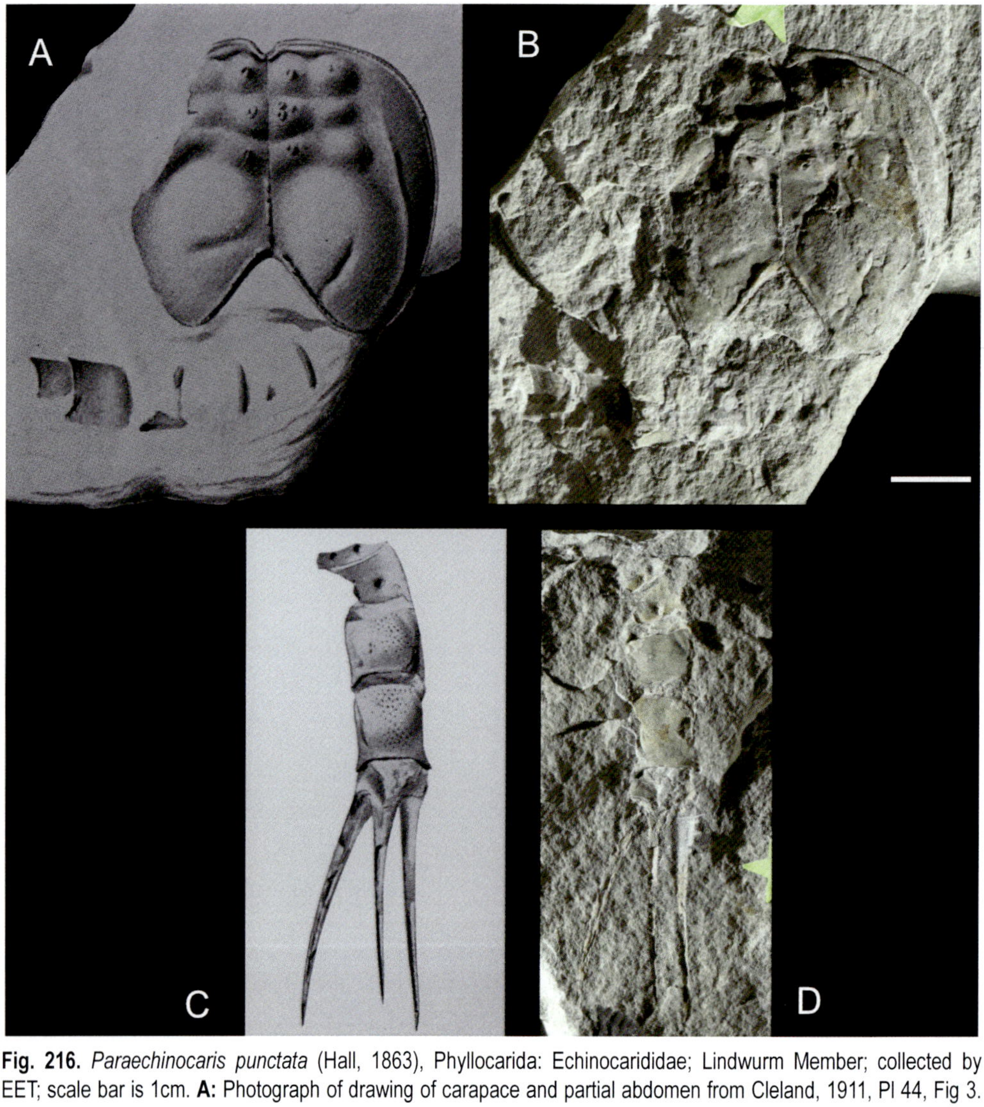

Fig. 216. *Paraechinocaris punctata* (Hall, 1863), Phyllocarida: Echinocarididae; Lindwurm Member; collected by EET; scale bar is 1cm. **A:** Photograph of drawing of carapace and partial abdomen from Cleland, 1911, Pl 44, Fig 3. Courtesy of Wisconsin Geological and Natural History Survey. **B:** PAL 78216a; the specimen on which Cleland based the drawing. Courtesy of Smithsonian Institution, Department of Paleobiology, National Museum of Natural History. **C:** Photograph of drawing of abdomen with telson from Cleland, 1911, Pl 44, Fig 4. Courtesy of Wisconsin Geological and Natural History Survey. **D:** PAL 78216b; the specimen on which Cleland based the drawing. Courtesy of Smithsonian Institution, Department of Paleobiology, National Museum of Natural History.

Chapter 17 — Crinoids, Blastoids, and Edrioasteroids
(Echinodermata)

The Echinodermata (Cambrian–Holocene) is the deuterostome animal phylum of nonsegmented, marine, benthic (exceptionally pelagic) metazoans lacking a differentiated head or brain and having radial (typically pentamerous) symmetry, a water vascular system, and an endoskeleton formed of plates and other distinct ossicles composed of crystalline calcite; includes the familiar starfish and brittle stars (Asterozoa), sea urchins, sand dollars, and sea cucumbers (Echinozoa), sea lilies and related forms (Crinozoa), and lesser known living and extinct subphyla (Ubaghs, 1967).

Each of the echinoderm species currently known from the Milwaukee Formation was from one of the following three subphyla: Crinozoa, Blastozoa, or Echinozoa.

Crinozoa (Cambrian–Holocene)
Crinozoans typically possess a calyx having ambulacra that serve as food grooves extending outward onto projecting appendages, which are parts of the body wall with coelomic extensions (arms or brachia) with ambulacral grooves lined with tube feet. A stalk (column) is typically attached to the aboral end of the calyx and possesses a disk-like or root-like holdfast at its distal end with which it attaches to the substrate.

Crinoidea (Ordovician–Holocene)
The Crinoidea is a class of crinozoans with pentamerous symmetry and five (or multiples of five) outspread, branched, food-gathering arms (brachia) (Ubaghs et al., 1978) (Fig. 217). Crinoid ossicles, especially columnals and sections of columns (Fig. 218), are some of the most abundant fossils in certain layers of the Milwaukee Formation.

The Crinoidea is generally considered to consist of five subclasses. All of the known crinoid species from the Milwaukee Formation are from the subclasses Camerata and Flexibilia, which are easily distinguished from each other. One of the obvious differences is that the camerates had a fairly well-defined calyx with rigidly connected plates and distinct branched arms with pinnules, while the flexible crinoids' calyx was not as well defined and arms are dichotomously branched and relatively broad, lacking pinnules.

At least six camerate crinoid species and subspecies from the Milwaukee Formation have been identified to date—all currently or tentatively assigned to *Melocrinites* (Figs. 219–235). The calyces of this genus and other camerates were on occasion the targets for commensalitic platyceratid gastropods that cemented their shells near the crinoid's anus, presumably to feed on the excrement of the host. Another notable point about *Melocrinites* is that each of its five arms is actually a brachial pair composed of two arms that are longitudinally "fused" (Fig. 236).

Taxocrinus telleri Springer, 1920, is to date the only species from the Flexibilia recognized from the Milwaukee Formation (Figs. 237 & 238). All *Taxocrinus* specimens have been found in a single layer representing a sediment smothering event in the Lindwurm Member. *Melocrinites* and blastoids of the Milwaukee Formation are not found in this layer. The *Taxocrinus* found at this horizon were living attached to a thin limestone layer which was abruptly covered by a clay rich thin layer of sediment that killed all of the crinoids. All or most of the specimens are preserved as scattered but commonly complete specimens with attachment, stem, and calyx intact. Because of difficulties in collecting complete specimens, none have been preserved in collections. Nearly all specimens of Milwaukee Formation calyces were collected from the Lindwurm Member, with the single exception of an internal mold of what may be a Greene collection *Melocrinites* from the upper Berthelet Member.

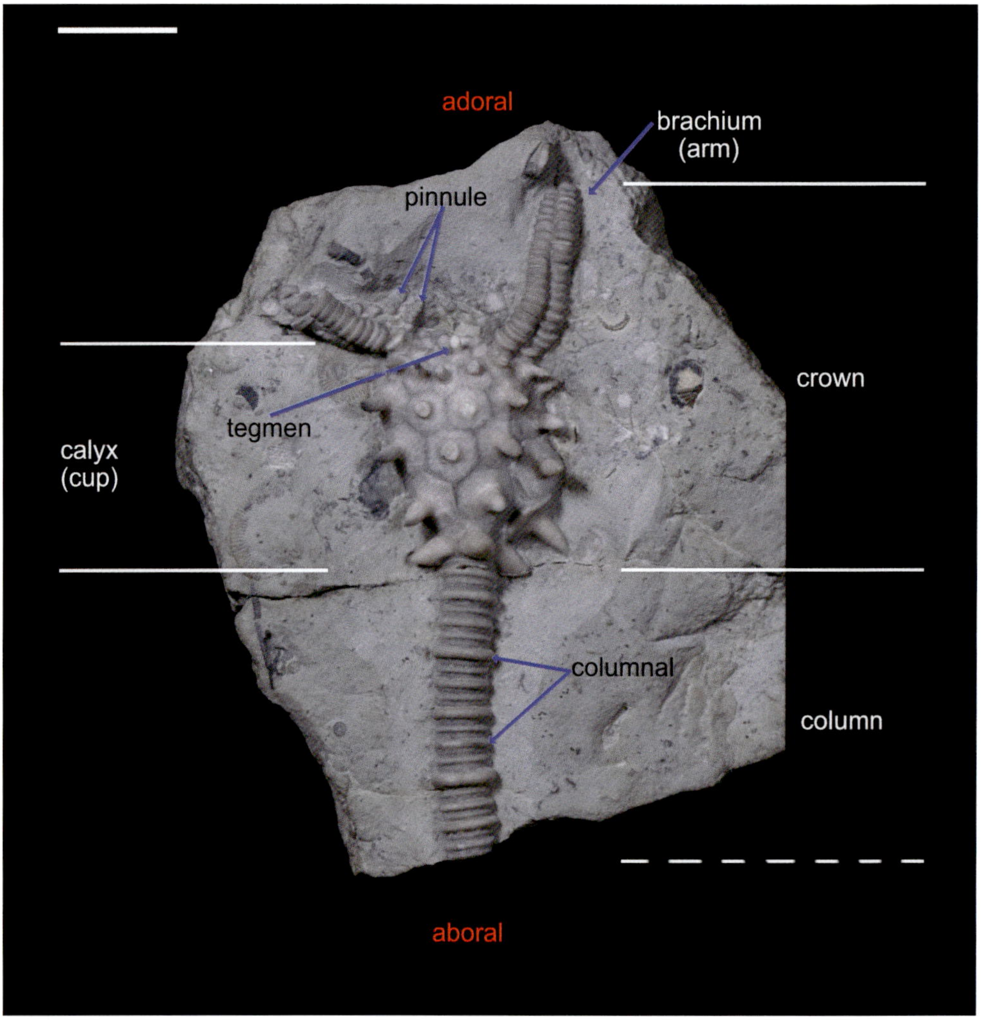

Fig. 217. The crinoid *Melocrinites nodosus spinosus* (Weller, 1898) with selected external morphological features and orientational terminology labeled; Lindwurm Member; MPM P28742; scale bar 1cm.

Fig. 218. Long section of column from *Taxocrinus*, and scattered crinoid ossicles, including a typical, circular columnal near bottom-right; also present are pieces of brachiopods, bryozoans, and corals; MPM P28170; Lindwurm Member; collected by KCG; scale bar 1cm.

Fig. 220. Portion of crinoid column; MPM 15369; Lindwurm Member; collected by GOR; scale bar 1cm.

Fig. 219. Crinoid column, probably from *Melocrinites*, Monobathrida: Melocrinitidae; MPM 15368; Lindwurm Member; collected by GOR; scale bar 5mm.

Fig. 221. Holdfast (center of photo) and portion of column from unidentified crinozoan; MPM 25375; Lindwurm Member; collected by MOE; scale bar 5mm.

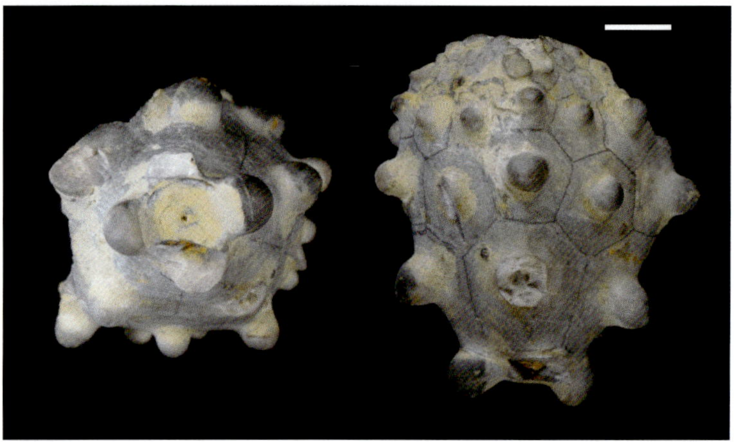

Fig. 222. Adoral and lateral views of *Melocrinites pentangularis* (Cleland, 1911), Monobathrida: Melocrinitidae; PAL 78289; holotype; Lindwurm Member; figured in Cleland, 1911, Pl 3, Fig 10; collected by EET; scale bar 5mm. Courtesy of Smithsonian Institution, Department of Paleobiology, National Museum of Natural History.

Fig. 223. Lateral view of calyx of *Melocrinites pentangularis* (Cleland, 1911), Monobathrida: Melocrinitidae; MPM 15431; Lindwurm Member; collected by GOR; scale bar 5mm.

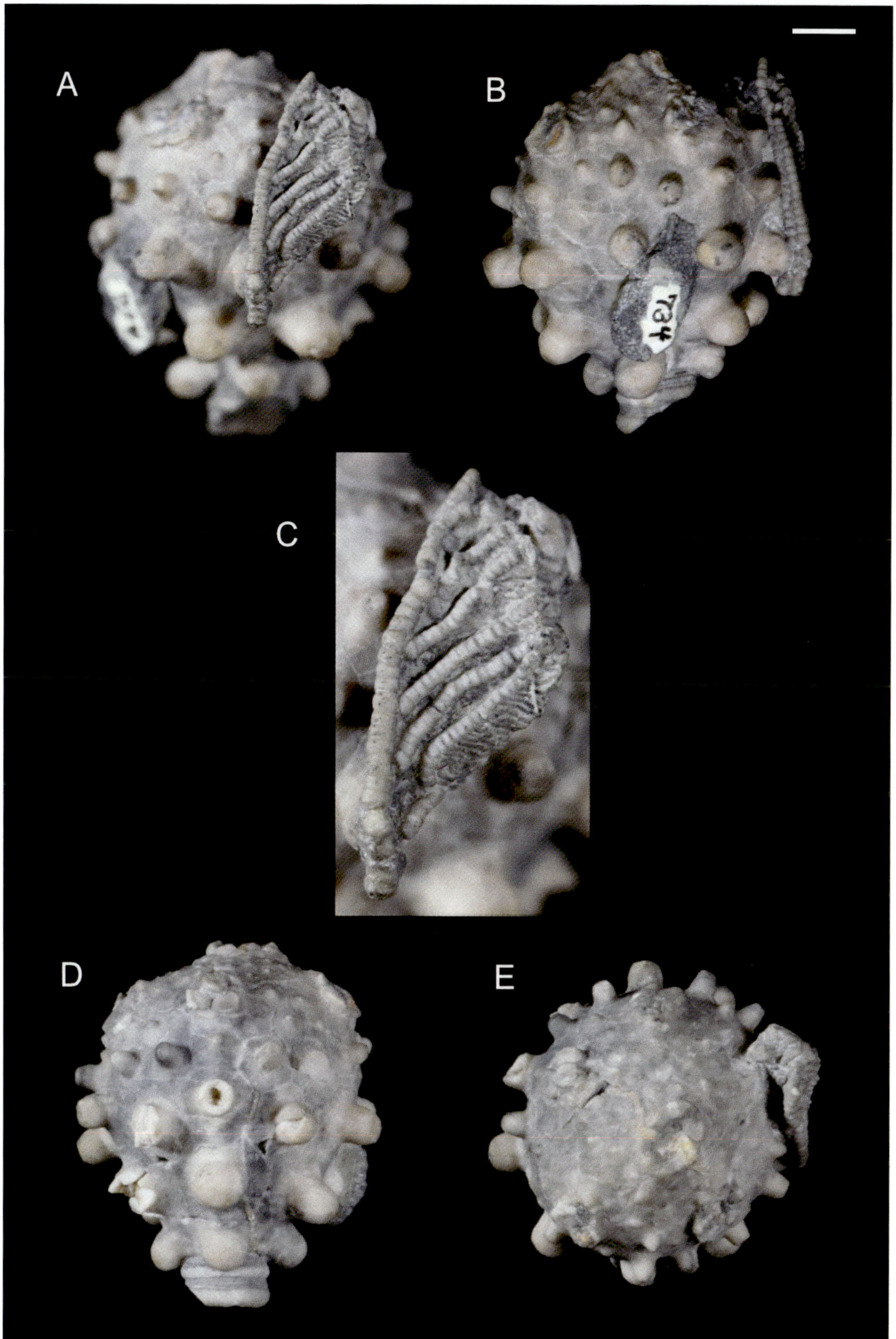

Fig. 224. Four views and a close-up of *Melocrinites pentangularis?* (Cleland, 1911), Monobathrida: Melocrinitidae; WESM-KCG 784; Lindwurm Member; collected by KCG; scale bar 5mm. **A:** Lateral view of calyx with a portion of a detached arm. **B:** A view showing the longitudinally fused structure of its double arm. **C:** Close-up of the the arm showing the fine projections arising from the bottom pinnule. **D:** This lateral view shows a small portion of the stalk. **E:** Aboral view. Prepared by David Comfort.

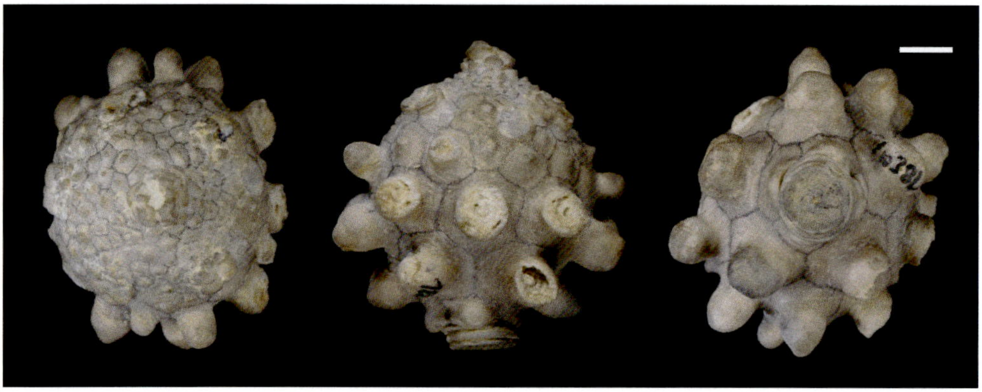

Fig. 225. Aboral, lateral and adoral views of *Melocrinites subglobosus* (Weller, 1898), Monobathrida: Melocrinitidae; PAL 78293; holotype; Lindwurm Member; figured in Weller, 1898, Pl 14, Fig 1; collected by EET; scale bar 5mm. Courtesy of Smithsonian Institution, Department of Paleobiology, National Museum of Natural History.

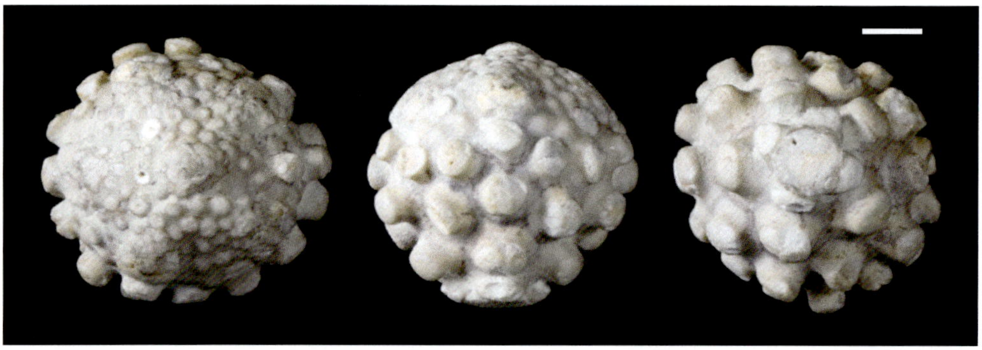

Fig. 226. Aboral, lateral, and adoral views of the calyx of *Melocrinites subglobosus* (Weller, 1898), Monobathrida: Melocrinitidae; WESM-KCG 19; Lindwurm Member; collected by KCG; scale bar 5mm.

Fig. 227. Aboral, lateral and adoral views of *Melocrinites milwaukeensis* (Weller, 1898), Monobathrida: Melocrinitidae; PAL 78296; holotype; Lindwurm Member; figured in Weller, 1898, Pl 14, Fig 5; collected by EET; scale bar 5mm. Courtesy of Smithsonian Institution, Department of Paleobiology, National Museum of Natural History.

Fig. 228. Aboral, lateral and adoral views of *Melocrinites milwaukeensis rotundus* (Weller, 1898), Monobathrida: Melocrinitidae; PAL 78295; holotype; Lindwurm Member; figured in Weller, 1898, Pl 14, Fig 4; collected by EET; scale bar 5mm. Courtesy of Smithsonian Institution, Department of Paleobiology, National Museum of Natural History.

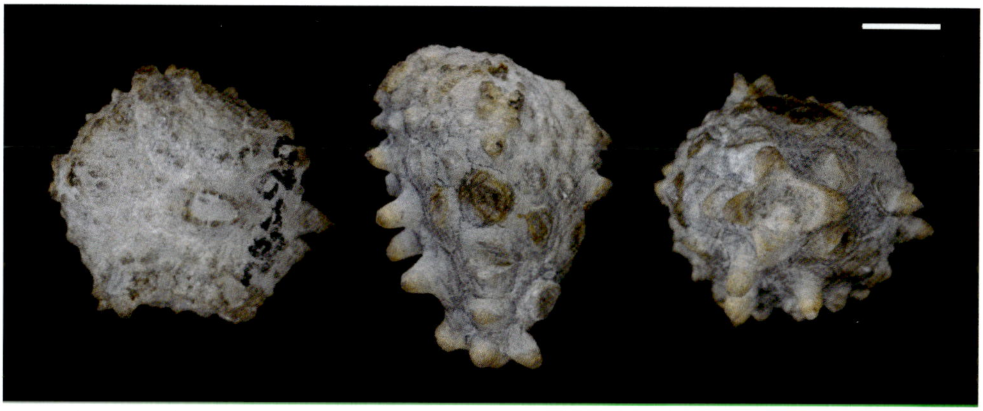

Fig. 229. Aboral, lateral and adoral views of *Melocrinites nodosus spinosus* (Weller, 1898), Monobathrida: Melocrinitidae; PAL 78290; holotype; Lindwurm Member; figured in Weller, 1898, Pl 14, Fig 2; collected by EET; scale bar 5mm. Courtesy of Smithsonian Institution, Department of Paleobiology, National Museum of Natural History.

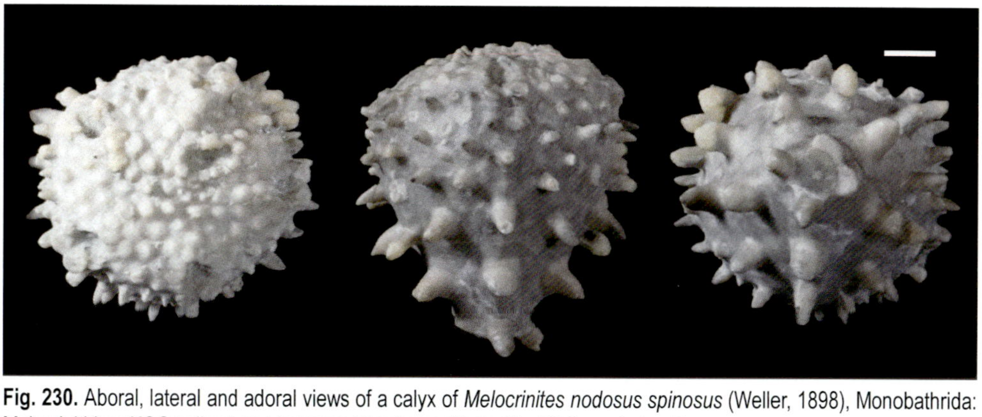

Fig. 230. Aboral, lateral and adoral views of a calyx of *Melocrinites nodosus spinosus* (Weller, 1898), Monobathrida: Melocrinitidae; KCG collection; Lindwurm Member; collected by KCG; scale bar 5mm.

Fig. 231. Lateral and adoral views of *Melocrinites nodosus* (Hall, 1861), Monobathrida: Melocrinitidae; MPM 15417; Lindwurm Member; collected by GOR; scale bar 5mm.

Fig. 232. Lateral view of calyx of *Melocrinites nodosus spinosus* (Weller, 1898), Monobathrida: Melocrinitidae; MPM 25848; Lindwurm Member; collected by DGM; scale bar 5mm.

Fig. 233. *Melocrinites nodosus spinosus* (Weller, 1898), Monobathrida: Melocrinitidae; calyx with portions of the stalk and two brachia attached; MPM P28742; Lindwurm Member; collected by KCG; scale bar 1cm; prepared by David Comfort.

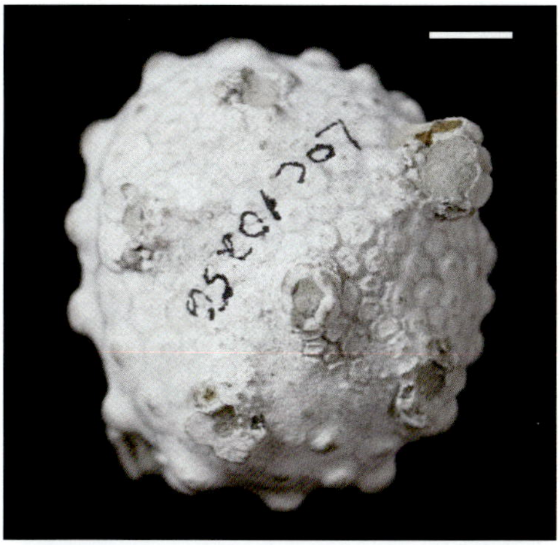

Fig. 234. Aboral view of *Melocrinites* sp. calyx clearly showing positions of the five brachia and the centrally-located anal tube; MPM 15419; Lindwurm Member; collected by GOR; scale bar 5mm.

Fig. 235. Two well preserved arms of an indeterminate crinoid, probably a camerate such as *Melocrinites*. This fossil was found in the glacial drift in Milwaukee. It is therefore attributed to the Milwaukee Formation with some reservation. MPM P28191; collected by SH; scale bar 1cm.

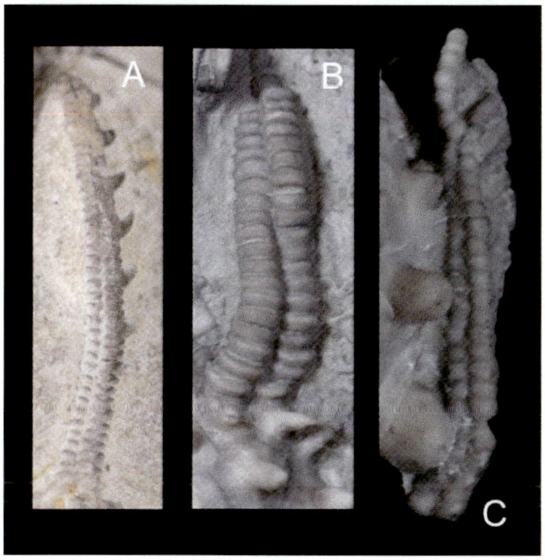

Fig. 236. Portions of the characteristic double arm of what appear to be three species of the crinoid *Melocrinites*. **A:** *M.?* sp.; MPM P28191. **B:** *Melocrinites nodosus spinosus* (Weller, 1898); WESM-KCG 784. **C:** *M. pentangularis* (Cleland, 1911); MPM P28742.

Fig. 237. Two lateral views of *Taxocrinus telleri* Springer, 1920, Taxocrinida: Taxocrinidae; PAL 78288; holotype; Lindwurm Member; figured in Springer, 1920, Pl 52, Figs 7a&b and Cleland, 1911, Pl 3, Fig 11; collected by EET; scale bar 5mm. Courtesy of Smithsonian Institution, Department of Paleobiology, National Museum of Natural History.

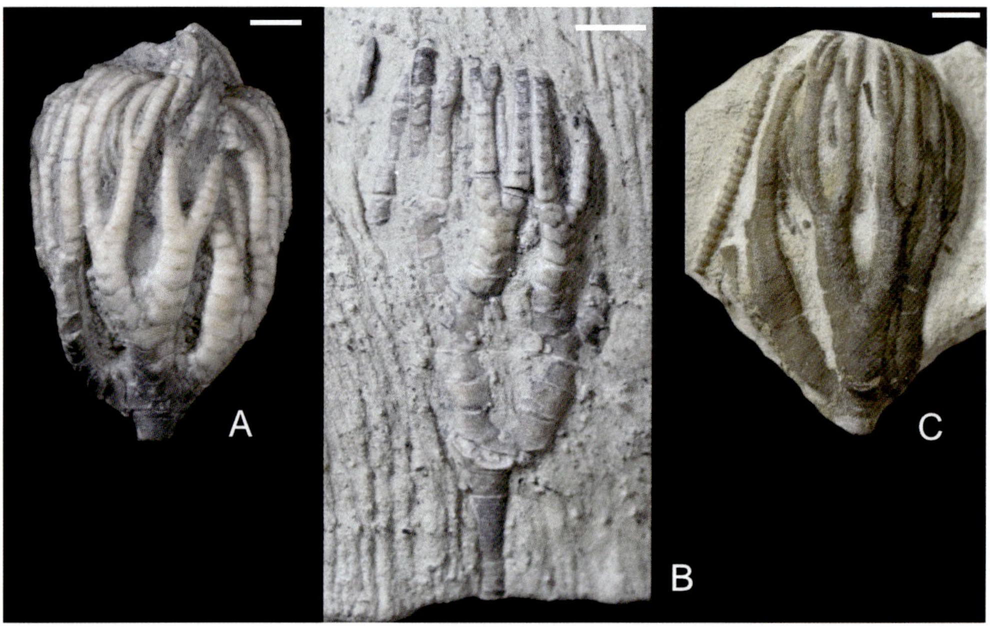

Fig. 238. *Taxocrinus telleri* Springer, 1920, Taxocrinida: Taxocrinidae; Lindwurm Member; scale bar 5mm. **A:** Calyx; KCG collection uncatalogued; collected by KCG; prepared by David Comfort. **B:** Calyx with part of stalk attached; MPM 25464. **C:** Calyx of *T. telleri* with a section of stalk that may also be from that species; MPM 25372; collected by MOE.

Blastozoa (Cambrian–Permian)

Blastozoans possessed a theca with ambulacra typically recumbent along the sides and possessed slender, biserial armlets termed brachioles, perhaps lacking tube feet, as well as specialized respiratory structures. Most were attached by a small holdfast and elevated on a column composed of round columnals.

Blastoidea (Silurian–Permian)

Blastoids were stalked echinoderms with biserial brachioles bordering their ambulacral areas, well developed pentameral symmetry, seventeen major plates arranged in three circlets, a lancet plate underlying its ambulacral system, an anus, mouth, infolded thin-walled calcareous linear structures (hydrospires) on either side of each ambulacrum crossing the plate suture, and a spiracle near the adoral extremity of each ambulacrum (Fay & Wanner, 1967) (Fig. 239). Three blastoid species were assigned previously to the Lindwurm Member, and additional species now appear to be present. Restudy of present material is needed due to incomplete preparation of old specimens and the discovery of new ones (Figs. 240–246).

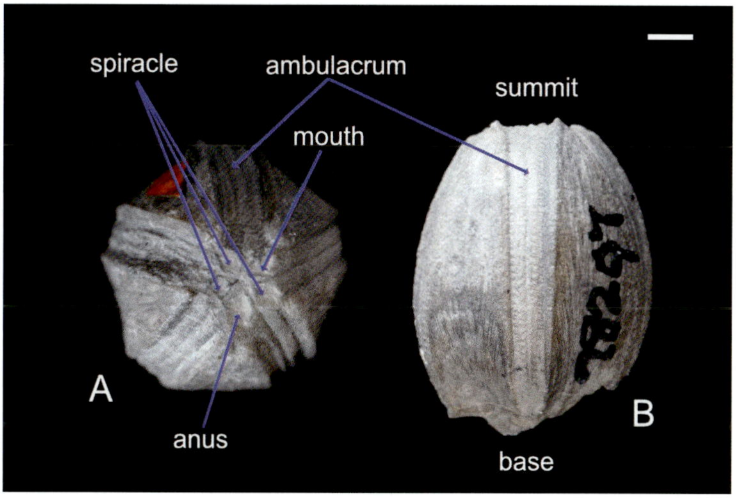

Fig. 239. Theca of the blastoid *Hyperoblastus milwaukeensis* (Weller, 1898) with selected external morphological features labeled; USNM PAL 78297; holotype; Lindwurm Member; figured in Weller, 1898, Pl 14, Fig 5; collected by EET; scale bar 2mm. **A:** Oral view. **B:** Lateral view. Courtesy of Smithsonian Institution, Department of Paleobiology, National Museum of Natural History.

Fig. 240. Partial theca of *Hyperoblastus* sp., Fissiculata: Astrocrinidae; MPM 284; Lindwurm Member; figured; collected by CEM; scale bar 2mm; identified as *Pentremitidea* cf. *P. milwaukeensis* in Cleland, 1911, Pl 3, Fig 1.

Fig. 241. *Nucleocrinus obovatus* (Barris, 1883), Spiraculata: Nucleocrinidae; MPM 5282; Lindwurm Member; figured in Cleland, 1911, Pl 3, Fig 2; scale bar 5mm.

Fig. 242. Three blastoid thecae; Lindwurm Member; scale bars 2mm. **A:** *Hyperoblastus milwaukeensis?* (Weller, 1898), Fissiculata: Astrocrinidae; MPM 15460; collected by GOR. **B:** *Hyperoblastus milwaukeensis?* (Weller, 1898); MPM 15462; collected by GOR. **C:** *Eleutherocrinus?* sp.; UWM-GM 5405.

Fig. 243. Two thecae of *Hyperoblastus* sp., Fissiculata: Astrocrinidae; Lindwurm Member; scale bar 2mm. **A:** MPM P28171; collected by KCG. **B:** MPM 25370; collected by MOE.

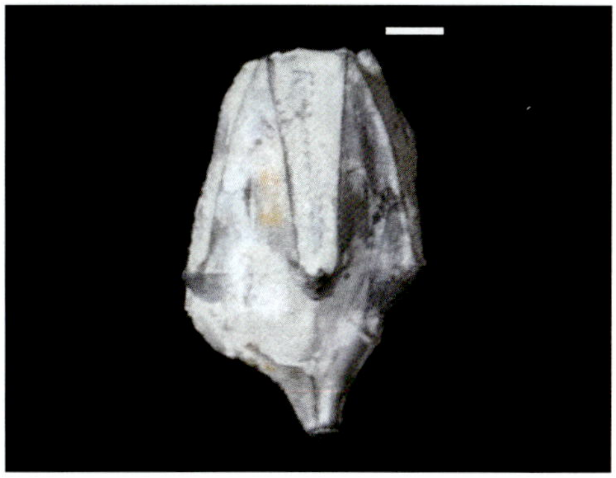

Fig. 244. Theca of *Hyperoblastus* cf. *H. filosus* (Whiteaves, 1891), Fissiculata: Astrocrinidae; MPM 285; Lindwurm Member; collected by CEM; scale bar 2mm. This theca is similar in general form to that from the Milwaukee Formation identified by Weller (1898, Pl 14, Fig 3) as *Pentremitidea filosa* (now reassigned to *Hyperoblastus*). Its greatest width is at the base of the ambulacra, as is characteristic of that species (Hoare, 1960).

Fig. 245. Two blastoid thecae and several disarticulating columnals from a coiled crinoid column (at left); Lindwurm Member; scale bar 2mm. **A:** *Hyperoblastus?* sp., Fissiculata: Astrocrinidae; MPM P25369A. **B:** *Hyperoblastus* sp.; MPM P25369B.

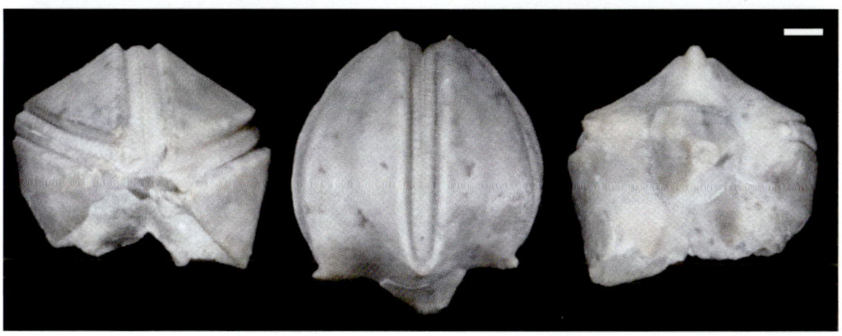

Fig. 246. Aboral, lateral, and adoral views of theca of *Hyperoblastus* sp., Fissiculata: Phaenoschismatidae; WESM-KCG 890; Lindwurm Member; collected by KCG; scale bar 2mm; prepared by David Comfort. This specimen appears to represent a species not previously recorded from the Milwaukee Formation.

Echinozoa (Cambrian–Holocene)

Echinozoans are echinoderms with radial symmetry, meridional pattern of growth producing an essentially globoid body, but no projecting appendages. Among the seven classes recognized at the time of writing this book, are the extinct helicoplacoids and edrioasteroids, and the extant echinoids (sea urchins and sand dollars).

Edrioasteroidea (Cambrian–Pennsylvanian)

Edrioasteroids were sedentary, stalkless, radiate echinozoans that had a domal, clavate, or globose polyplated theca; ambulacra (generally five) endothecal, formed by floor plates and cover plates (Ubaghs, 1967). In the Lindwurm Member edrioasteroids have been found attached to the brachiopod *Devonatrypa?* (Figs. 247 & 248). Unfortunately, the ambulacra and other diagnostic features of the specimens found thus far are poorly known due to disassociation of the thecal ossicles, thus preventing precise identification. Based on examination of several specimens, Colin Sumrall (personal communication, 04-Dec-2017) suggested that this taxon is probably *Krama* or perhaps *Agelacrinites*. Edrioasteroids were not previously recorded from the Devonian of Wisconsin.

Fig. 247. Four edrioasteroids, possibly *Krama* sp., attached to valves of the atrypid brachiopod *Devonatrypa?*; Lindwurm Member; scale bars 5mm. **A:** Only the attachment rim remains on this specimen; MPM 16199; collected by GOR. **B:** A better preserved specimen, but still no distinguishable ambulacra; MPM 15471; collected by GOR. **C:** Edrioasteroid reduced to a jumble of ossicles; MPM P28192; collected by HR?; prepared by David Comfort. **D:** The only specimen known to retain the ambulacra. This is the only photograph of this specimen. It is from a photo proof sheet prepared by Joseph Emielity at MPM that includes images of some of the other edrioasteroids shown in this book. The image is numbered 2G251. The scale was not given on the photograph. The specimen cannot be located.

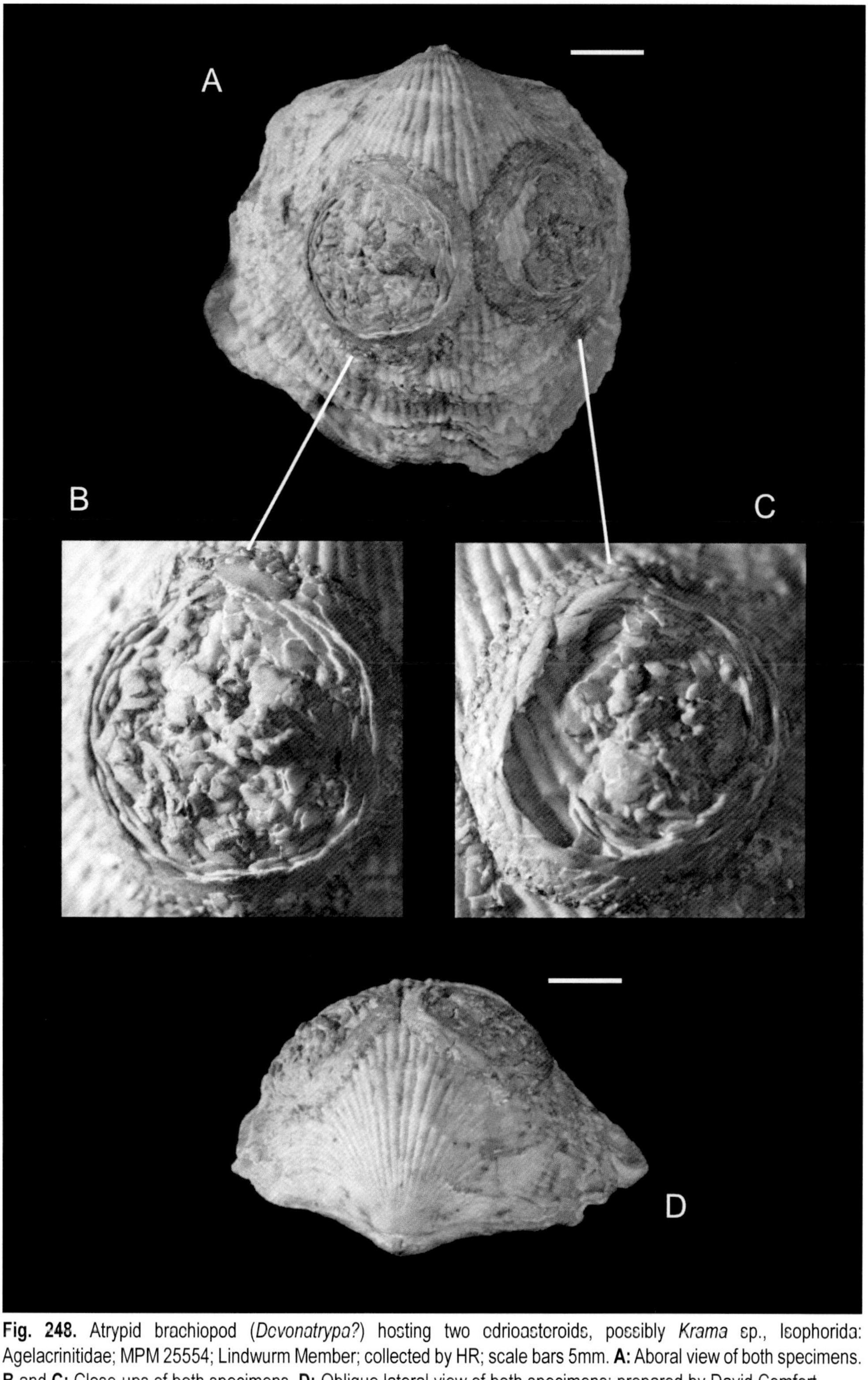

Fig. 248. Atrypid brachiopod (*Devonatrypa?*) hosting two edrioasteroids, possibly *Krama* sp., Isophorida: Agelacrinitidae; MPM 25554; Lindwurm Member; collected by HR; scale bars 5mm. **A:** Aboral view of both specimens. **B** and **C:** Close-ups of both specimens. **D:** Oblique lateral view of both specimens; prepared by David Comfort.

Chapter 18 — Dendroid Graptolites
(Hemichordata: Graptolithina: Dendroidea)

The Graptolithina (Cambrian–Pennsylvanian) was a class of sessile or pelagic colonial marine hemichordates that secreted a sclerotized exoskeleton with characteristic growth bands and growth lines; the thecae housing individual zooids were usually arranged in a single or double row along the branches (stipes) of the colony (rhabdosome); rhabdosomes originated by a single bud from the initial zooid, housed in a conical sicula, producing simple, branched or rarely encrusting colonies (Bulman, 1970). Graptolites are usually considered to be a class of the phylum Hemichordata along with the classes Enteropneusta, Pterobranchia, and Planctoshaeroidea.

The Dendroidea (Cambrian–Pennsylvanian) was an order of sessile graptolites attached by the apex of the sicula, which was then generally more or less embedded in secondary cortical tissue forming a root-like base; stipes composed of stolothecae, autothecae, and bithecae were produced by regular triad budding. The rhabdosome was typically erect, dendroid in habit of growth (Bulman, 1970). All of the graptolite fossils found in the Milwaukee Formation to date are dendroids.

Prior to publication of this volume, graptolites were unrecorded from the Devonian rocks of Wisconsin. Their presence is still known from only a few rhabdosomes found in the Lindwurm Member at Estabrook Park. Based on images of some of this material, David Loydell recognized *Callograptus*, *Dendrograptus*, and *Desmograptus* (personal communication, 03-Sep-2017). Those images are shown here (Fig. 249).

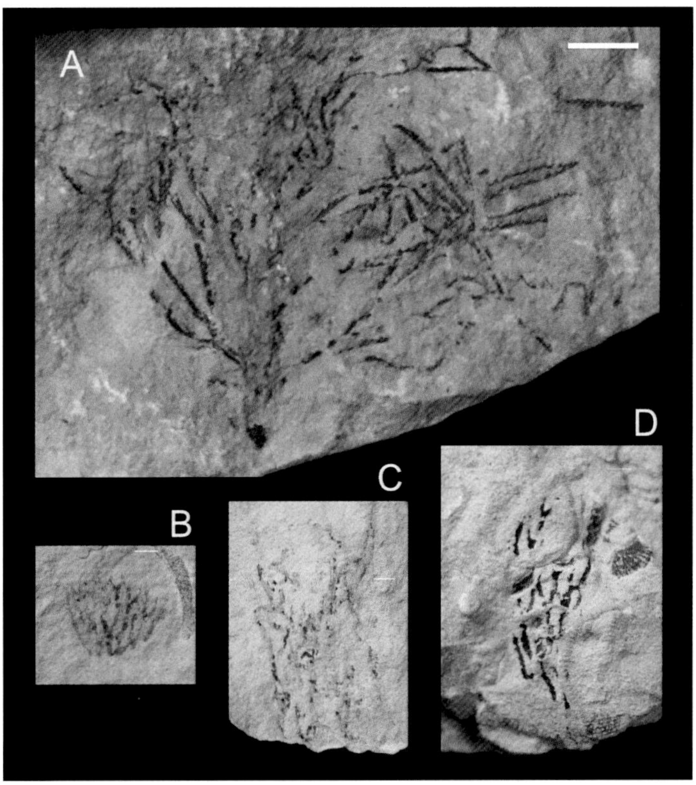

Fig. 249. Dendroid graptolites, Dendroidea: Dendrograptidae; Lindwurm Member. **A:** *Dendrograptus* sp.; MPM 22972; collected by BB; scale bar 5mm. **B:** *Desmograptus* sp.; MPM-P28172; collected by LAG; scale bar 2mm; **C:** insufficiently preserved for identification; MPM P28173; collected by KCG; scale bar 2mm. **D:** *Callograptus* sp.; MPM P28174; collected by KCG; scale bar 2mm.

Chapter 19 — Conodonts
(Chordata: Conodontomorpha: Conodonta)

The Conodonta (Cambrian–Triassic) was a class of small, marine, pelagic and benthic chordates of eel-like form having head regions containing assemblages of microscopic phosphatic tooth-like structures also termed conodonts (Fig. 250). The conodont assemblages are generally considered to be feeding apparatuses. Enigmatic for centuries until fossils of its body were finally discovered, conodonts bear many of the characteristics of the chordates. Due to evidence of cellular bone, eyes, wear patterns on the elements, etc., some researchers have placed them within the Craniata if not the vertebrata itself (Donoghue et al., 2000). However, further findings including no evidence of a distinct cranium, suggest placement of the conodonts in their own subphylum alongside the cephalochordates and craniates (Turner et al., 2010; Nelson et al., 2016).

Schumacher (1971a,b) identified about twenty "species" of conodont elements from the Lindwurm and Berthelet Members; the North Point Member was not sampled for his study (Figs. 251 & 252). Because a conodont animal had several "species" of conodont elements in its assemblage, the number of conodont animal species represented in these figures is probably less than twenty.

Mayer (2001) stated that the upper one meter of the Milwaukee Formation includes *Icriodus subterminus* Youngquist, 1947, *Polygnathus dengleri* Bischoff & Ziegler, 1957, *P. dubius* Hinde, 1879, and *Mehlina gradata* Youngquist, 1945. In the same abstract he also reported that below that section, presumably in the North Point Member, *Icriodus subterminus* and *Polygnathus ovatinodosus* Ziegler & Klapper, 1976, were found.

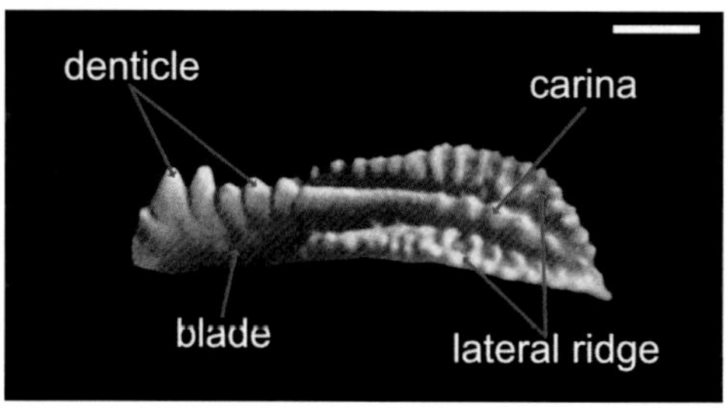

Fig. 250. Upper lateral view of the conodont element *Polygnathus pennatus* Hinde, 1879 with selected external morphological features labeled; UW-793 A; scale bar 0.2mm. Courtesy of D. Schumacher and Wisconsin Geological and Natural History Survey.

Fig. 251 (next page). Conodont elements; collected by DS; scale bar 0.2mm; all images courtesy of D. Schumacher and Wisconsin Geological and Natural History Survey. **A:** *Icriodus nodosus* (Huddle, 1934); upper view; UW-691; figured in Schumacher, 1971, Pl 9, Fig 20. **B:** *Acodina* sp.; posterior view; UW-628; figured in Schumacher, 1971c, Pl 7, Fig 3. **C:** *Elsonella rhenana* Lindström & Ziegler, 1965; lateral view; UW-659; figured in Schumacher, 1971c, Pl 7, Fig 4. **D:** *Hindeodella austinensis* Stauffer, 1940; inner lateral view; UW-664 A; figured in Schumacher, 1971c, Pl 8, Fig 2. **E:** *Neoprioniodus pronus* (Huddle, 1934); lateral view; UW-706 B; figured in Schumacher, 1971c, Pl 7, Fig 8. **F:** *Synprioniodina forsenta* Stauffer, 1940; lateral view; UW-800; figured in Schumacher, 1971c, Pl 7, Fig 6. **G:** *Hindeodella austinensis* Stauffer, 1940; inner lateral view; UW-646; figured in Schumacher, 1971c, Pl 8, Fig 12. **H:** *Polygnathus decorosus* Stauffer, 1938; upper lateral view; UW-770; figured in Schumacher, 1971c, Pl 11, Fig 10. **I:** *Hibbardella* sp.; posterior view; UW-661; figured in Schumacher, 1971c, Pl 7, Fig 13. **J:** *Prioniodina aversa* Stauffer, 1940; lateral view; UW-798; figured in Schumacher, 1971c, Pl 7, Fig 14. **K:** *Ligonodina panderi* (Hinde, 1879); inner lateral view; UW-693 A; figured in Schumacher, 1971c, Pl 7, Fig 15. **L:** *Hindeodella priscilla* Stauffer, 1938; inner lateral view; UW-665 A; figured in Schumacher, 1971c, Pl 8, Fig 4. **M:** *Ozarkodina* sp.; lateral view; UW-709; figured in Schumacher, 1971c, Pl 7, Fig 17.

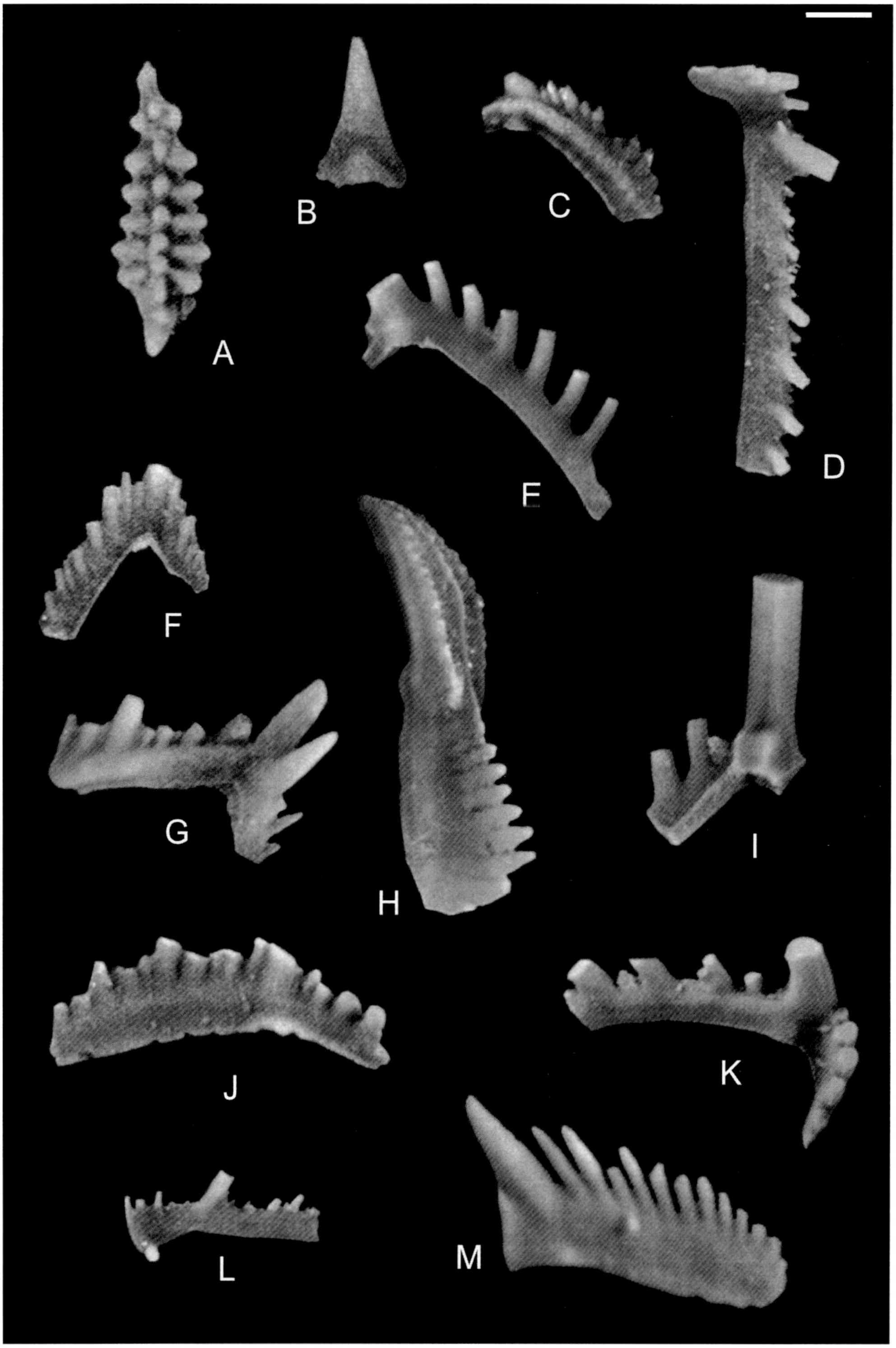
A
B
C
D
E
F
G
H
I
J
K
L
M

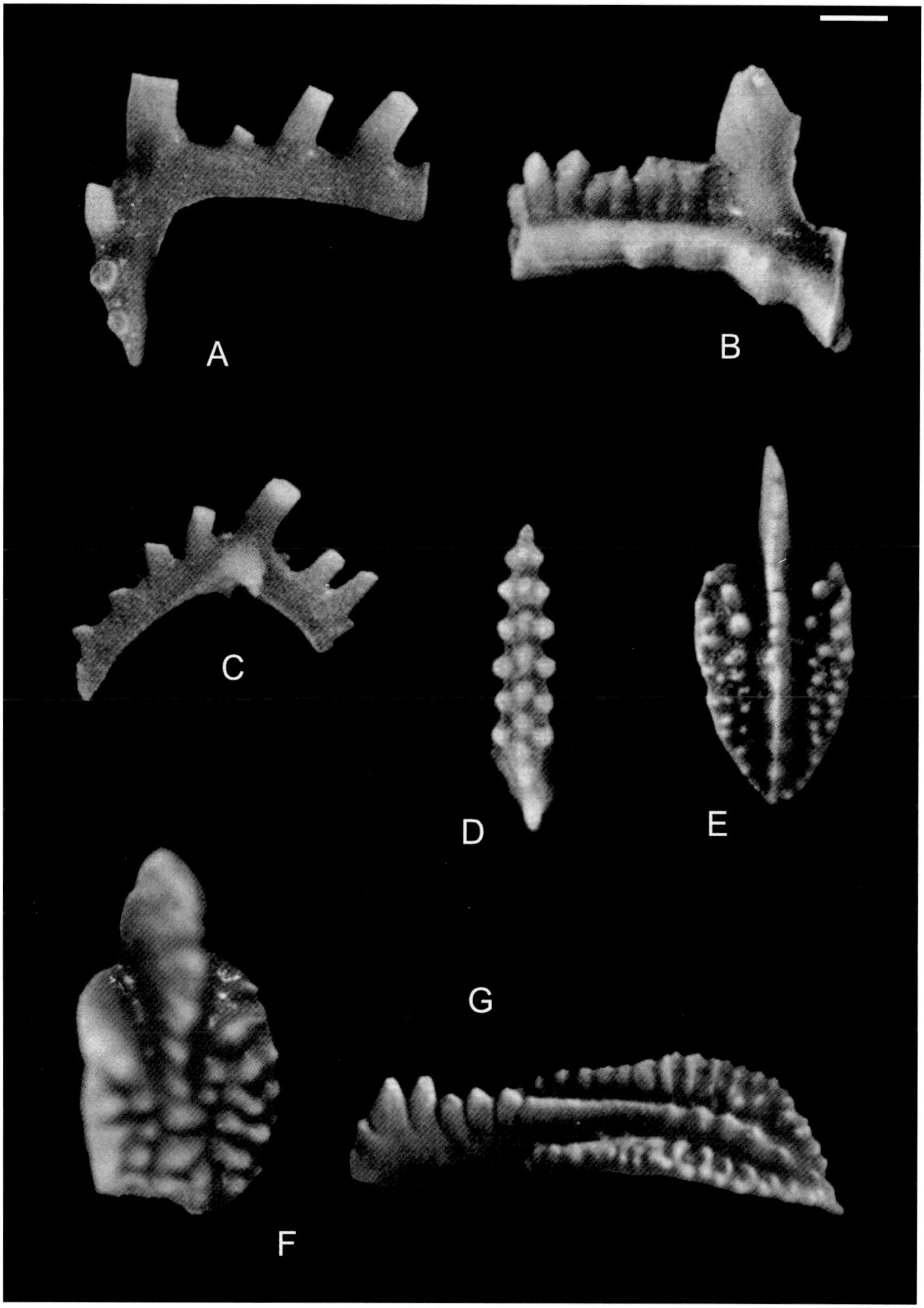

Fig. 252. Conodont elements; collected by DS; scale bar 0.2mm; all images courtesy of D. Schumacher and Wisconsin Geological and Natural History Survey. **A:** *Ligonodina panderi* (Hinde, 1879); inner lateral view; UW-693; figured in Schumacher, 1971c, Pl 8, Fig 5. **B:** *Bryantodus* cf. *B. colligatus* (Bryant, 1921); lateral view; UW-659; figured in Schumacher, 1971c, Pl 8, Fig 6. **C:** *Lonchodina* cf. *L. arcuata* Ulrich & Bassler, 1926; inner lateral view; UW-699; figured in Schumacher, 1971c, Pl 8, Fig 7. **D:** *Icriodus* sp.; upper view; UW-688; figured in Schumacher, 1971c, Pl 9, Fig 31. **E:** *Polygnathus cristatus* Hinde, 1879; upper view; UW-790 A; figured in Schumacher, 1971c, Pl 10, Fig 1. **F:** *Polygnathus ordinatus* Bryant, 1921; upper view; UW-790; figured in Schumacher, 1971c, Pl 10, Fig 9. **G:** *Polygnathus pennatus* Hinde, 1879; upper lateral view; UW-793 A; figured in Schumacher, 1971c, Pl 10, Fig 13.

Chapter 20 — Fishes

What is a fish? Taxonomists still use this term, but do not clearly define it. Readers will notice that the authors agree with the classification scheme of Nelson (1994) in which Pisces is no longer recognized as a formal taxonomic group, in order to better reflect the phylogenetic relationships revealed by fossils, comparative anatomy and molecular genetic evidence now at hand. The reason is essentially that Pisces was always defined to exclude the tetrapods; however, sarcopterygians (lobe-finned fish) are now known to be more closely related to tetrapods (including humans) than to the rest of the "fishes." As a result, most taxonomists now place humans and the other tetrapods within the Sarcopterygia itself, and have abandoned Pisces, rather than to either perpetuate what is thus known to be a paraphyletic group or to completely re-define it to include tetrapods. An inconvenience that comes along with this new classification is that in effect no more "Pisces" means no more "fishes." What we do have are several living and extinct infraphyla of craniates that most people would agree range from "clearly fish" to "fish-like." One of these, the hagfish (infraphylum Myxinomorphi), is not even a vertebrate. Then there are the fish-like conodonts (see previous chapter) that almost make it to fishhood but appear to lack a cranium. The point is that there is currently no formal taxonomic group called "fishes." The authors therefore take the liberty of referring to all taxa discussed in the remaining sections of this chapter as fishes.

As mentioned earlier, it was the fossil fishes that brought attention to Devonian rocks in Wisconsin and pointed to their age. This is somewhat ironic, given that fish fossils in the Milwaukee Formation are relatively rare (Cleland, 1911), consisting mostly of isolated plates, teeth, spines, and scales. Complete skeletons have yet to be found. Nonetheless, the fish fauna of this formation is considerably diverse, as shown by more than twenty species distributed among four classes: Placodermi, Chondrichthyes, Acanthodii, and Osteichthyes. These classes are discussed in the next four sections.

Placodermi (Silurian–Devonian)

The placoderms were a class of marine and freshwater gnathostome (jawed) craniates having a mostly cartilaginous endoskeleton and boney dermal plates covering their head and trunk (Figs. 253–255). Arising early in the jawed fish lineage, placoderms may have given rise to some of the other major groups of jawed fish (Zhu et al., 2013), although some paleontologists consider them an evolutionary dead end. They were the dominant fish in the Devonian Period, although they became extinct at its end, and they included some of the largest predators known from the Paleozoic Era (Nelson et. al., 2016). The class is best known for its giant predators, but also includes filter feeders and smaller species and forms whose teeth suggest shell crushing or scavenging lifestyles. Three (possibly four) of its six orders are represented in the Milwaukee Formation.

<u>Arthrodiriformes (Devonian)</u>

This was the most familiar and diverse placoderm order. Perhaps the most famous arthrodire was the well-known fearsome predator, *Dunkleosteus terrelli* (Newberry, 1873), from the Cleveland Shale of Ohio, which was up to six meters long. The predatory arthrodire species from the Milwaukee Formation were nearly as large, attaining a length of about five meters (Boylan, 1973, p. 324) in the case of Milwaukee's most common arthrodire, *Eastmanosteus pustulosus* (Eastman, 1897) (Figs. 256–274). *E. pustulosus* is the type species of that genus. The type specimen of that species is from the Milwaukee Formation. Plates that appear to be from its smaller relative, *E. tuberculatus* (Newberry, 1889), have been found in a Milwaukee Cement Company quarry (Cleland, 1911), as well as plates of what has been suggested to be a new tuberculated species of the arthrodire *Titanichthyes* (Eastman, 1898 & 1907). The later specimens are maintained in the Milwaukee Public Museum (e.g., Fig. 275) and in the National Museum of Natural History. As suggested by its name, *Titanichthyes* was a giant, attaining a length of seven meters. In contrast to its predatory relatives mentioned above, *Titanichthyes* was probably a filter-feeder, as indicated by a lack of sharp cusps on its jaws (Long, 2011, p. 88). Fossils from three other genera often referred to arthrodires, *Eczematolepis* (Figs. 276–278), *Holonema* (Fig. 279), and *Ostophorus* (based on PAL 328013, NMNH V

14964, and four other fin spines residing at NMNH) also inhabited Milwaukee's Devonian sea. Hussakof & Bryant (1918, p. 102) suggested that the fin spines assigned to the former might instead belong to the ptyctodont *Palaeomylus*, the dental plates of which are sometimes associated with them. The present authors herein tentatively refer *Eczematolepis* to the Arthrodiriformes.

Precise identification of isolated dermal plates from arthrodires can be difficult or impossible; however, careful examination of bone structure and ornamentation can lead to more reliable assignments in some cases (Fig. 280).

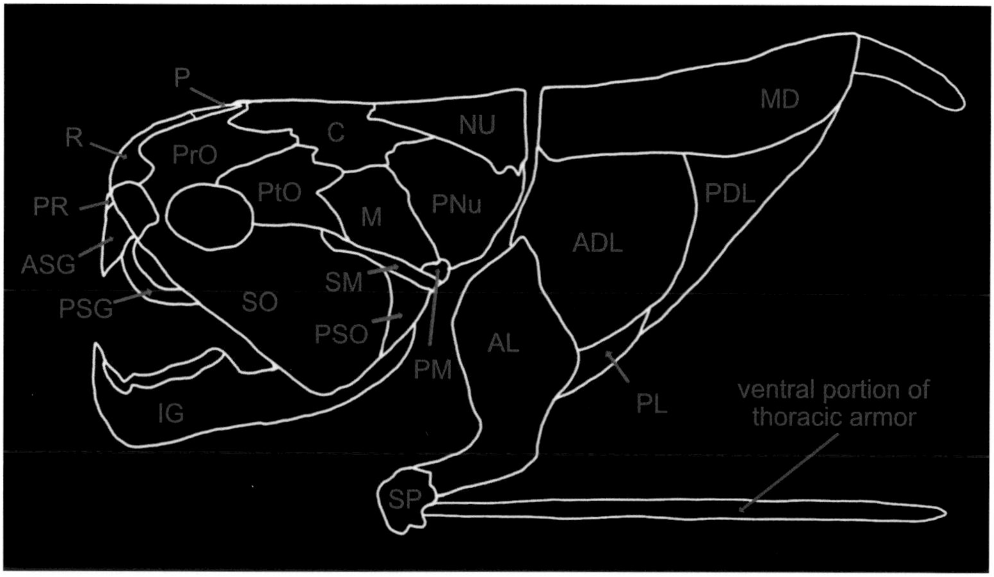

Fig. 253. Lateral view of the dermal armor of *Eastmanosteus pustulosus* (Eastman, 1897) with dermal plates labeled; after Boylan, 1975, Fig. 62.

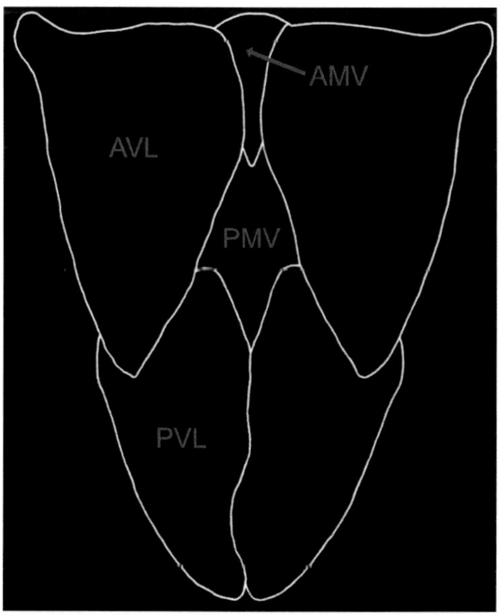

Fig. 254. Ventral portion of the thoracic armor of *Eastmanosteus pustulosus* (Eastman, 1897) with dermal plates labeled; after Boylan, 1975, Fig. 64.

ADL	anterodorsolateral
AL	anterolateral
AMV	anteromedianventral
ASG	anterosupragnathal
AVL	anteroventrolateral
C	central
IG	infragnathal
M	marginal
MD	mediandorsal
NU	nuchal
P	pineal
PDL	posterodorsolateral
PL	posterolateral
PM	postmarginal
PMV	posteromedianventra
PNu	paranuchal
PR	prerostral
PrO	preorbital
PSG	posterosupergnathal
PSO	postsuborbital
PTO	postorbital
PVL	posteroventrolateral
R	rostral
SM	submarginal
SO	suborbital
SP	spinal

Fig. 255. Acronyms for dermal bone elements used in this book.

Fig. 256. Left ADL plate of *Eastmanosteus pustulosus* (Eastman, 1897), Arthrodiriformes: Dunkleosteidae; VPF 1381; holotype; Berthelet Member; figured in Eastman, 1897, Pl 3, Fig 4; collected by FHD; scale bar 1cm. Courtesy of Museum of Comparative Zoology, Harvard University. ©President and Fellows of Harvard College.

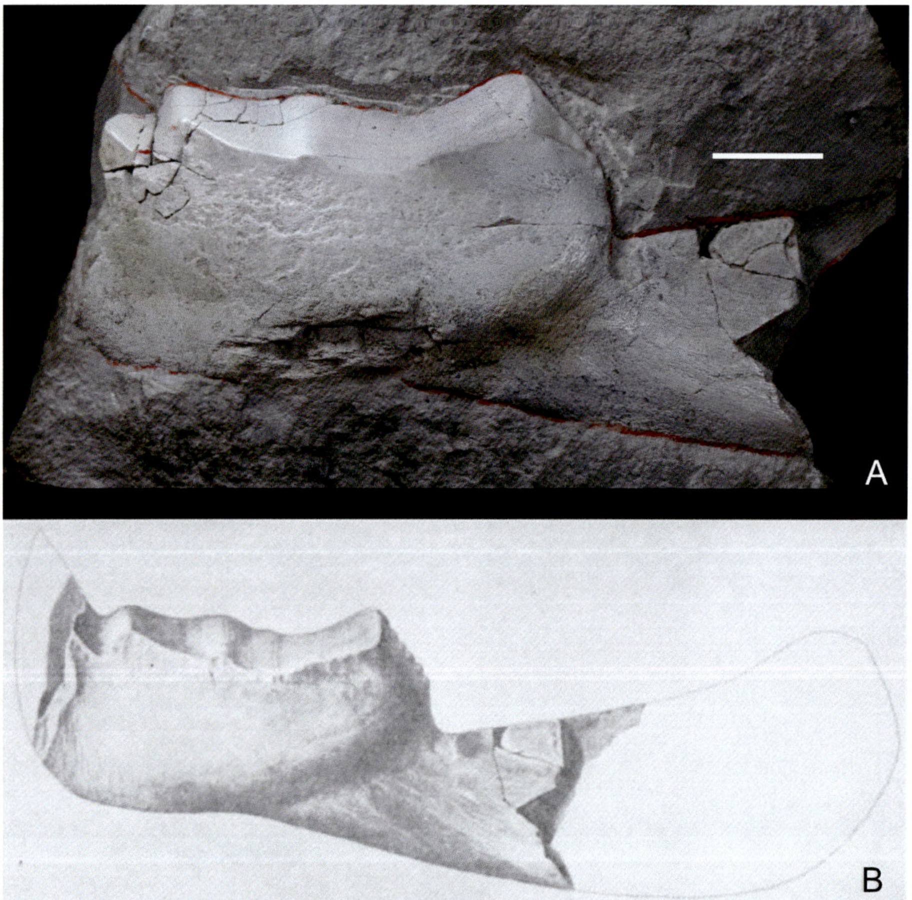

Fig. 257. IG plate ("lower jaw") of *Eastmanosteus pustulosus* (Eastman, 1897), Arthrodiriformes: Dunkleosteidae; Berthelet Member, figured in Eastman, 1900, Fig 1, Cleland, 1911, Pl 45, Fig 1, and Boylan, 1973, pages 409 and 410; collected by CEM. **A:** Photograph of the fossil; scale bar 2cm. **B:** Eastman's (1900) drawing of the fossil with a reconstruction of the anterior and posterior extremities.

Fig. 258. PSG of *Eastmanosteus pustulosus* (Eastman, 1897), Arthrodiriformes: Dunkleosteidae; UWM-GM 10000; Berthelet Member; figured in Boylan, 1973, Fig 30b; scale bar 1cm.

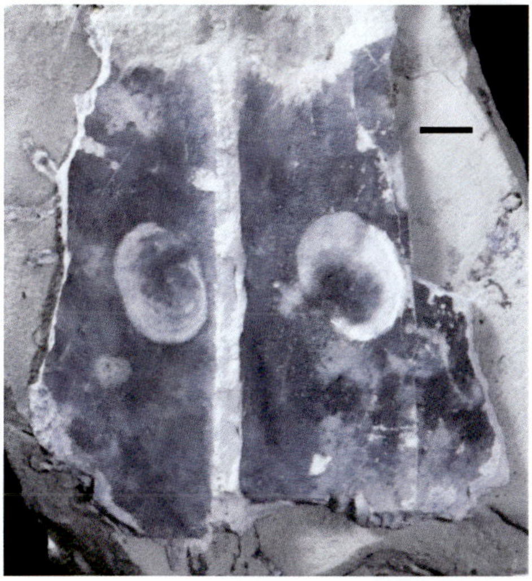

Fig. 259. Portion of unidentified dermal plate, possibly from *Eastmanosteus pustulosus* (Eastman, 1897), Arthrodiriformes: Dunkleosteidae; UWM-GM 25719; Berthelet Member; scale bar 1cm.

Fig. 260. Possible internal view of anterior medial ventral (AMV) plate of *Eastmanosteus pustulosus* (Eastman, 1897), Arthrodiriformes: Dunkleosteidae; UWM-GM 25803; Berthelet Member; scale bar 1cm.

Fig. 261. SO plate of *Eastmanosteus pustulosus* (Eastman, 1897), Arthrodiriformes: Dunkleosteidae; UWM-GM 10083; Berthelet Member; figured in Boylan, 1973, Figs 20 a&b; scale bar 1cm.

Fig. 262. Possible PVL of *Eastmanosteus pustulosus* (Eastman, 1987), Arthrodiriformes: Dunkleosteidae; UWM-GM 25807; Berthelet Member; scale bar 1cm.

Fig. 263. Dorsal and lateral views of possible MD plate of *Eastmanosteus pustulosus* (Eastman, 1987), Arthrodiriformes: Dunkleosteidae; UWM-GM 9837; Berthelet Member; scale bar 1cm.

Fig. 264. Possible MD plate of *Eastmanosteus pustulosus* (Eastman, 1897), Arthrodiriformes: Dunkleosteidae; UWM-GM 21005; Berthelet Member; scale bar 1cm.

Fig. 265. Partial skull roof of *Eastmanosteus pustulosus* (Eastman, 1897) with both central plates present, Arthrodiriformes: Dunkleosteidae; UWM-GM 21004; Berthelet Member; scale bar 1cm.

Fig. 266. Possible MD plate of *Eastmanosteus pustulosus* (Eastman, 1897), Arthrodiriformes: Dunkleosteidae; UWM-GM 693; Berthelet Member; collected by CEM; scale bar 1cm.

Fig. 267. Internal view of skull roof of *Eastmanosteus pustulosus* (Eastman, 1897), Arthrodiriformes: Dunkleosteidae; MPM 467 VP; collected by CEM; scale bar 1cm.

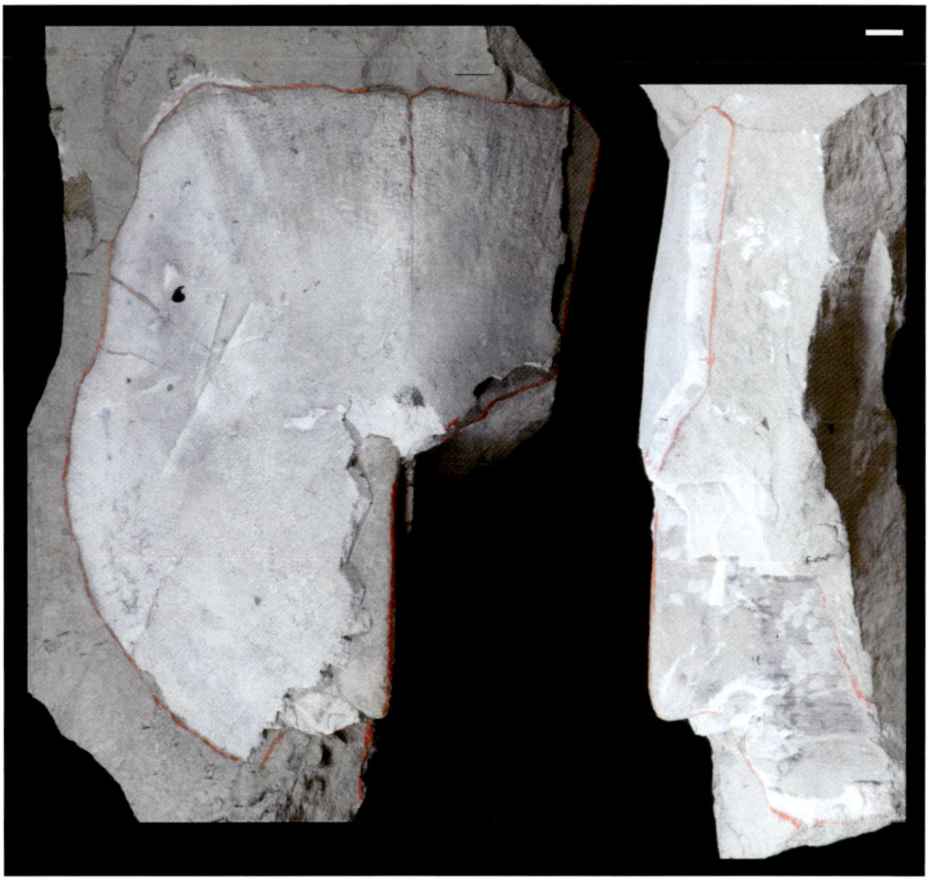

Fig. 268. Dorsal and lateral views of MD plate of *Eastmanosteus pustulosus* (Eastman, 1897), Arthrodiriformes: Dunkleosteidae; MPM 443; Berthelet Member; figured in Boylan, 1973, Figs 38 a&b; collected by CEM; scale bar 1cm.

Fig. 269. Dorsal and posterior views of MD plate of *Eastmanosteus pustulosus* (Eastman, 1897), Arthrodiriformes: Dunkleosteidae; MPM 712; Berthelet Member; collected by CEM; scale bar 1cm.

Fig. 270. Left PVL plate of *Eastmanosteus pustulosus* (Eastman, 1897), Arthrodiriformes: Dunkleosteidae; MPM 689; Berthelet Member; collected by CEM; scale bar 1cm.

Fig. 271. Five views of an unidentified boney structure labeled as *"Eastmanosteus pustulosus;"* MPM 478; Berthelet Member; collected by CEM; scale bar 1cm.

Fig. 272. Three views of ADL plate of *Eastmanosteus pustulosus* (Eastman, 1897), Arthrodiriformes: Dunkleosteidae; MPM 684; Berthelet Member; collected by SW; scale bar 1cm.

Fig. 273. Unidentified plate labeled as *"Eastmanosteus pustulosus;"* MPM 695; Berthelet Member; collected by CEM; scale bar 1cm.

Fig. 274. Interior view of dermal plate from *Eastmanosteus pustulosus* (Eastman, 1897), Arthrodiriformes: Dunkleosteidae; UWM-GM 10109; Berthelet Member; scale bar 1cm.

Fig. 275. Dermal plate attributed to a right dorso-lateral plate of *Titanichthyes* sp., Arthrodiriformes: Titanichthyidae; MPM 480; Berthelet Member; collected by CEM; scale bar 1cm.

Fig. 276. Oblique close-up view of fin spine of *Eczematolepis telleri* (Eastman, 1898), Arthrodiriformes ?: Phlyctaeniidae ?; V 14981; holotype; Berthelet Member; figured in Eastman, 1898, Fig 49; collected by EET; scale bar 5mm. Courtesy of Smithsonian Institution, Department of Paleobiology, National Museum of Natural History.

Fig. 277. Fin spine of *Eczematolepis telleri* (Eastman, 1898), Arthrodiriformes?: Phlyctaeniidae?; UWM-GM 18271; Berthelet Member; scale bar 1cm.

Fig. 278. Fin spine of *Eczematolepis telleri* (Eastman, 1898), Arthrodiriformes?: Phlyctaeniidae?; MPM 470; Berthelet Member; collected by CEM; scale bar 1cm.

Fig. 279. Dermal plate from *Holonema rugosum* (Claypole, 1886), Arthrodiriformes: Brachythoraciidae. Hussakof & Bryant (1918) stated that this specimen "is of great interest for showing on the outer, ornamented face several teeth marks apparently inflicted by a contemporary animal, probably a dinichthyid." V 14978; Berthelet Member; figured in Hussakof & Bryant, 1918, Pl 32, Fig 1; collected by EET; scale bar 5mm. Courtesy of Smithsonian Institution, Department of Paleobiology, National Museum of Natural History.

Fig. 280. Unidentified dermal plate; UWM-CM 10001; Berthelet Member; scale bar 1cm.

Ptyctodontiformes (Devonian)

This placoderm order was characterized by its sturdy, blunt, frequently textured chewing surfaces presumably specialized for crushing shells (Fig. 281). Its head and trunk region was not as heavily armored as that of the arthrodires. Male ptyctodonts had a pair of structures on their abdomens that may have been claspers used for mating. Dermal plates, spines and scales of ptyctodonts are the most common fish fossils from this formation. Four ptyctodont genera, *Ptyctodus*, *Rhynchodus*, *Gamphacanthus*, and *Palaeomylus*, have been identified from the Milwaukee Formation (Figs. 282–295).

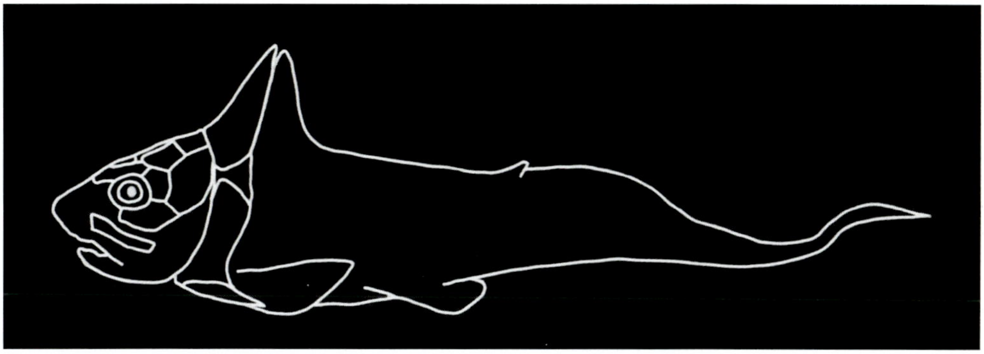

Fig. 281. Dermal plates, especially gnathal elements and fin spines, of these small, shell-crushing ptyctodont placoderms, are the most common fish fossils in the Milwaukee Formation. After Carr & Jackson, 2008, p. 3.

Fig. 282. Oblique view of left upper tooth plate of *Ptyctodus ferox* Eastman, 1898, Ptyctodontiformes: Ptyctodontidae; V 14966; paratype; Berthelet Member; figured in Eastman, 1898, Fig 35; collected by EET; scale bar 5mm. Courtesy of Smithsonian Institution, Department of Paleobiology, National Museum of Natural History.

Fig. 283. Seven ptyctodont tooth plates, probably *Pyctodus ferox* Eastman, 1898, Ptyctodontiformes: Ptyctodontidae; UWM-GM 10050; Berthelet Member; scale bar 1cm.

Fig. 284. Tooth plate of *Ptyctodus calceolus* (Newberry & Worthen, 1866), Ptyctodontiformes: Ptyctodontidae; V 14880; Berthelet Member; collected by EET; scale bar 5mm. Courtesy of Smithsonian Institution, Department of Paleobiology, National Museum of Natural History.

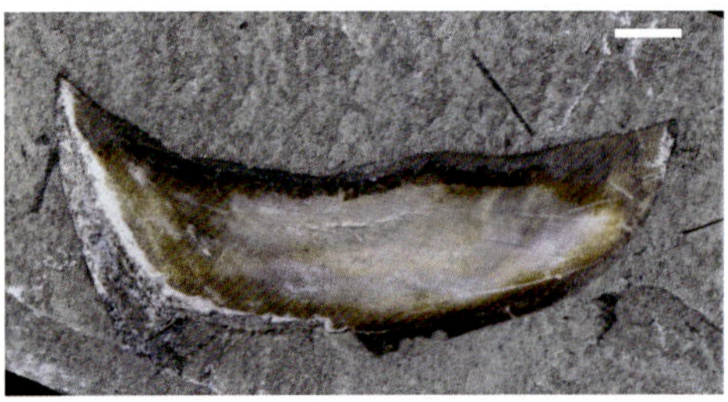

Fig. 285. Tooth plate from *Rhynchodus excavatus* Newberry, 1877, Ptyctodontiformes: Ptyctodontidae; V 14918; Berthelet Member; collected by EET; scale bar 5mm; Courtesy of Smithsonian Institution, Department of Paleobiology, National Museum of Natural History.

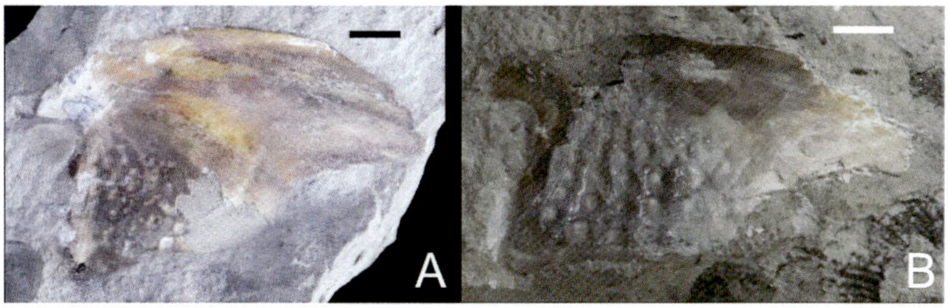

Fig. 286. *Rhynchodus* scales; Ptyctodontiformes: Ptyctodontidae; Berthelet Member; scale bars 5mm. **A:** UWM-GM 9985. **B:** MPM 484; collected by GOR.

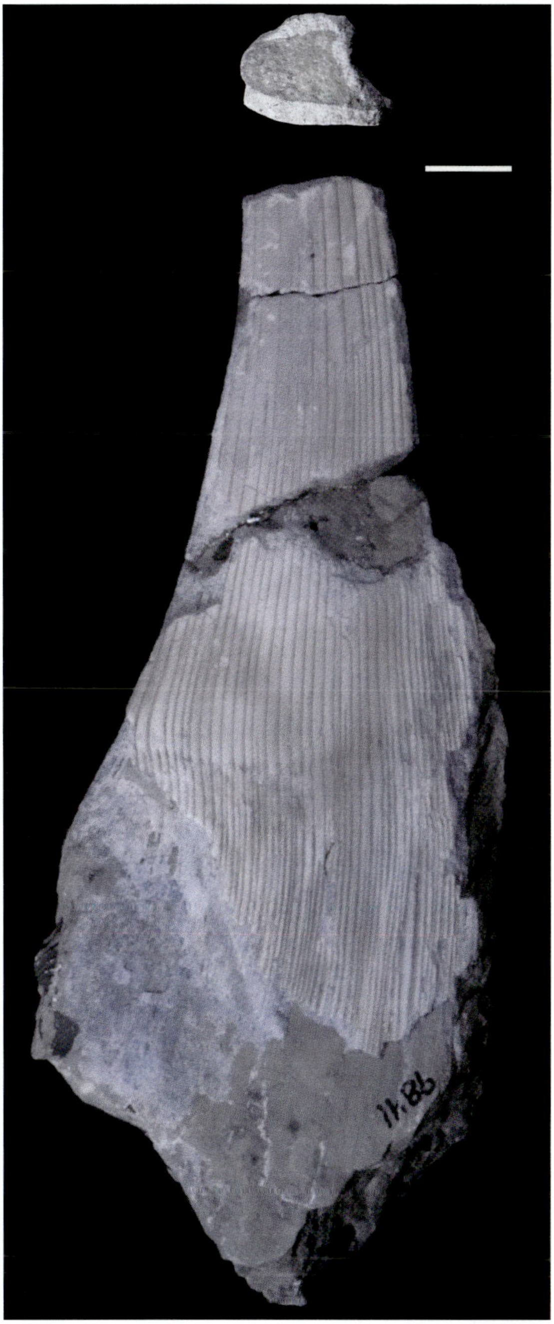

Fig. 287. Fin spine (cross sectional view shown at top) of *Gamphacanthus politus* (Newberry, 1889), Ptyctodontiformes: Ptyctodontidae; UWM-GM 9841; Berthelet Member; scale bar 1cm.

Fig. 288. Fin spine of *Gamphacanthus politus* (Newberry, 1889), Ptyctodontiformes: Ptyctodontidae; MPM 473; Berthelet Member; collected by CEM.

Fig. 289. Fin spine of *Gamphacanthus* sp., Ptyctodontiformes: Ptyctodontidae; MPM 723; Berthelet Member; collected by CEM; scale bar 5mm.

Fig. 290. Two pieces of a fin spine from *Gamphacanthus uddeni* (Lindahl, 1895), Ptyctodontiformes: Ptyctodontidae; V 14881; Berthelet Member; collected by EET; scale bar 5mm. Courtesy of Smithsonian Institution, Department of Paleobiology, National Museum of Natural History.

Fig. 291. Tooth plates from *Palaeomylus greenei* (Newberry, 1889), Ptyctodontiformes: Ptyctodontidae; Berthelet Member; collected by EET; scale bars 5mm. **A:** V 14934. **B:** PAL 328017. **C:** V 14877. Courtesy of Smithsonian Institution, Department of Paleobiology, National Museum of Natural History.

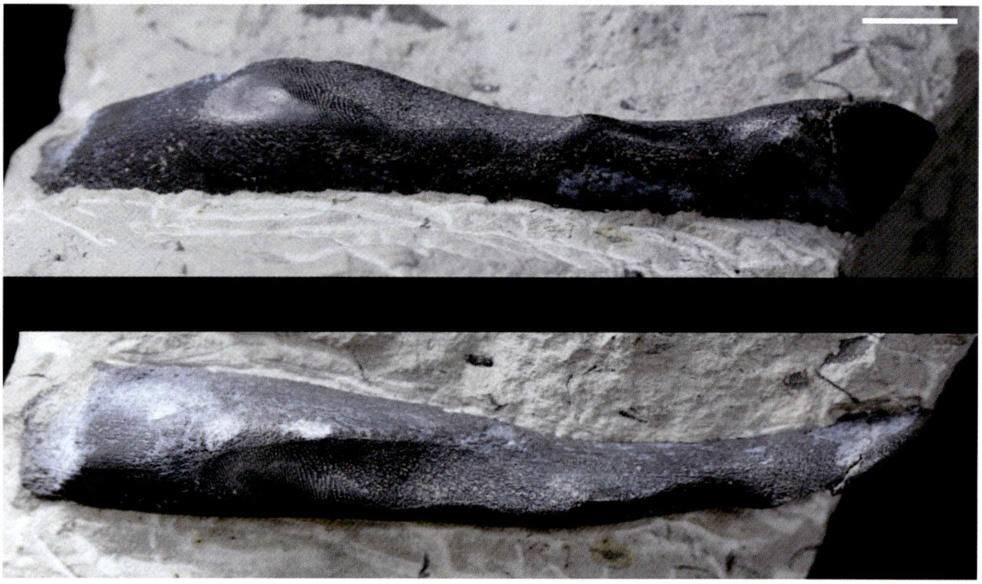

Fig. 292. Lateral and grinding surface views of a ptyctodont gnathal plate, Ptyctodontiformes: Ptyctodontidae; MPM 479; Berthelet Member; Milwaukee Cement Company; scale bar 5mm.

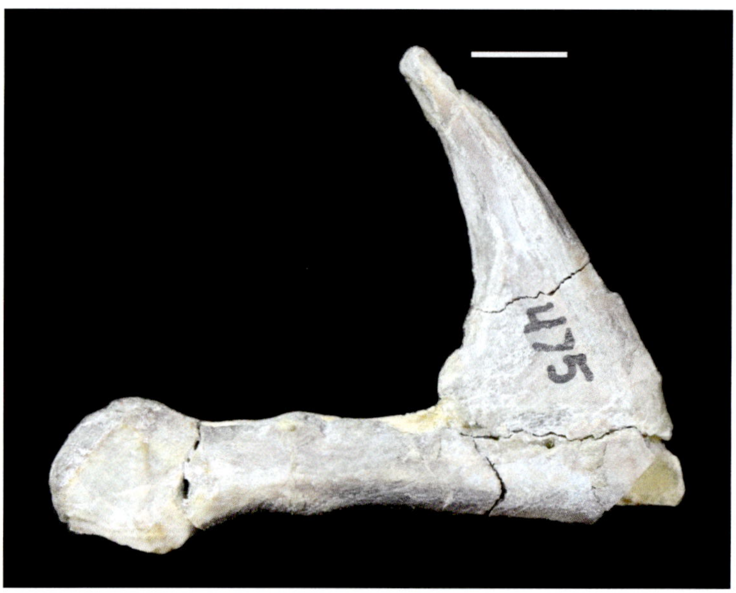

Fig. 293. Partial ptyctodont tooth plate, Ptyctodontiformes: Ptyctodontidae; MPM 475; Berthelet Member; collected by CEM; scale bar 5mm.

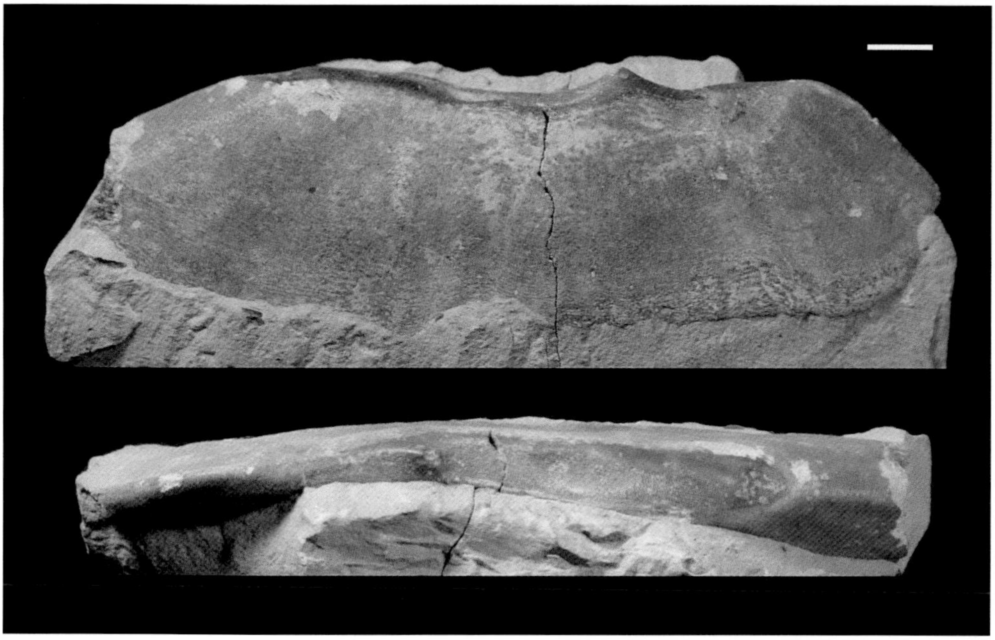

Fig. 294. Lateral and grinding surface views of ptyctodont gnathal plate, Ptyctodontiformes: Ptyctodontidae; UWM-GM 10238; Berthelet Member; scale bar 1cm.

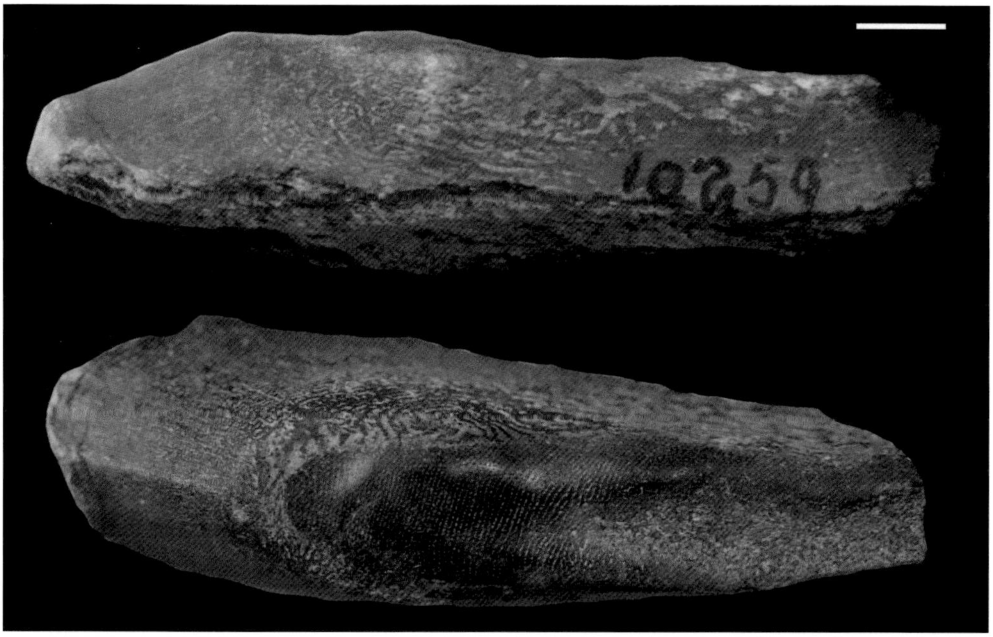

Fig. 295. Lateral and grinding surface views of ptyctodont gnathal plate, Ptyctodontiformes: Ptyctodontidae; UWM-GM 10259; Berthelet Member; scale bar 5mm.

Rheneniformes (Devonian)

This order included the distinctive Asterosteidae—a family of flattened, ray-like forms typified by the large, conoidal, stellate-based tubercles on its dermal plates. Its type genus, *Asterosteus*, has been found in the Milwaukee Formation (Fig. 296).

Fig. 296. Coarsely tuberculated dermal plate of *Asterosteus* sp., Rhenaniformes: Asterosteidae. The close-up shows the large, stellate-based conoidal tubercles typical of this genus. VPF 8523; Berthelet Member; collected by EET; scale bar 1cm. Courtesy of Museum of Comparative Zoology, Harvard University. ©President and Fellows of Harvard College.

Petalichthyiformes (Devonian)

This placoderm order consisted of small, presumably bottom-feeding, individuals whose dermal plates had rows of tubercles on their outer surfaces and thick tubes observable in the head region on the inside surfaces of the dermal plates (Fig. 297). One specimen (V14890) from the Teller collection has been identified as a skull element of the petalichthyid *Macropetalichthys*.

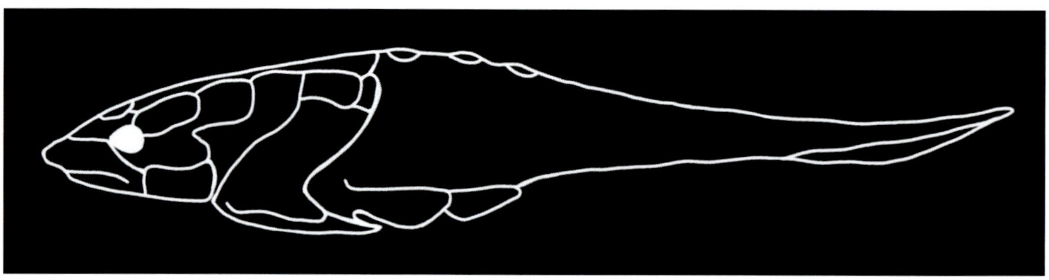

Fig. 297. A dermal plate believed to have been from the petalichthyiform placoderm *Macropetalichthys* was collected in the Milwaukee Formation. After Nelson et al., 2016, p. 39.

Chondrichthyes (Silurian–Holocene)
The Chondrichthyes is the class of gnathostomes that includes sharks, rays, chimaeras, and related forms. Like the placoderms and the other fish classes that were extant when sharks first appeared, their endoskeletons (except for a few elements such as teeth, spines and some scales) were composed of cartilage rather than bone. Like the ptyctodonts (and possibly other placoderms), the males have a pair of claspers which are used in mating.

While chondrichthyans would evolve a great variety of forms, sizes and lifestyles enabling them to live in many aquatic environments as predators, scavengers or filter feeders, the Devonian chondrichthyans, such as those found in the Milwaukee Formation, were relatively small and conservative in form.

Two of Milwaukee's known chondrichthyan species were known from only a few specimens (Fig. 298). *Cladodus monroei* Eastman, 1900, is a cladoselachid based on a 2.5cm-wide tooth from Milwaukee. Smaller specimens that may belong to this species were reportedly collected from the Lindwurm Member at Estabrook Park (J. Emielity, personal communication to K Gass, 1998). None of these *Cladodus* specimens can currently be located (Sumpter & Meyer, 1991). J. Maisey (personal communication, 17-Dec-2017) confirmed that a fin spine from the Berthelet Member (Fig. 299) is consistent with a ctenacanthid shark such as *Ctenacanthus*.

Cleland (1911) listed "*Myriacanthus*-like dermal plates" from the Berthelet Member. This would be the only known material of that genus from the Milwaukee Formation and the earliest occurrence of the chondrichthyan order **Chimaeriformes**, which is not otherwise known from before the Mississippian (Early Carboniferous); however, the whereabouts of this material is unknown. The identification of this material is suspect and needs to be verified should this material re-surface.

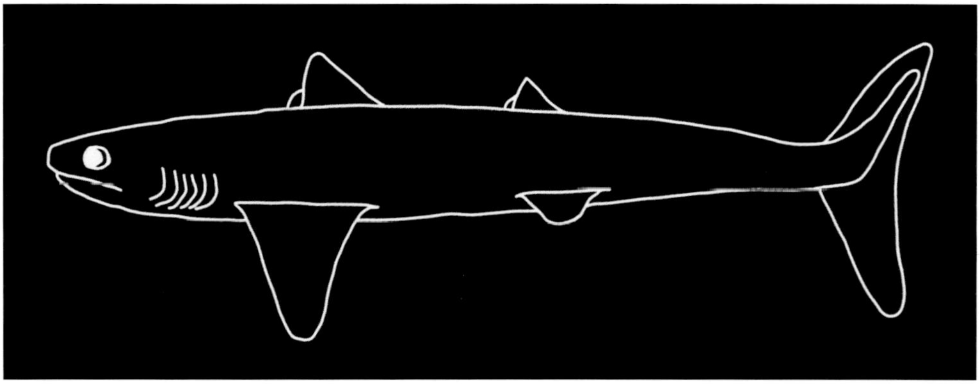

Fig. 298. The primitive shark *Cladodus* is known in the Milwaukee Formation only from its teeth. After Nelson et al., 2016, p. 46.

Acanthodii (Ordovician? or Silurian–Permian)
Acanthodians were small (less than 2.5 meters) gnathostomes with dermal and perichondral bone as well as a series of stout spines before the dorsal, anal and paired fins (Fig. 300). Certain anatomical features suggest that they were mid- and surface-water feeders (Nelson et al., 2016). Eastman (1907) indicated that some fragments from what is now known as the Milwaukee Formation suggest the acanthodian *Machaeracanthus longaevus*? Eastman, 1907 or *M. sulcatus* Newberry, 1889. One of those fragments (a dermal plate in two pieces) has been identified as *M. longaevus* and is shown here (Fig. 301). A fragment of a fin spine labelled as *Machaeracanthus* sp. resides at the Museum of Comparative Zoology (Fig. 302).

Fig. 299. Fin spine of *Ctenacanthus?* sp. or related shark, Ctenacanthiformes: Ctenacanthidae; PAL 328559; Berthelet Member; collected by EET; scale bar 5mm. Courtesy of Smithsonian Institution, Department of Paleobiology, National Museum of Natural History.

Fig. 301. Two dermal plate fragments of *Machaeracanthus longaevus* (Eastman, 1907), Ischnacanthaformes: Ischnacanthidae; V 14965; Berthelet Member; collected by EET; scale bar 5mm. The exact relation of these fragments to each other is uncertain. Courtesy of Smithsonian Institution, Department of Paleobiology, National Museum of Natural History.

Fig. 302. Portion of fin spine of *Machaeracanthus* sp., Ischnacanthaformes: Ischnacanthidae; VPF 13259; Berthelet Member; collected by FHD; scale bar 5mm. Courtesy of Museum of Comparative Zoology, Harvard University. ©President and Fellows of Harvard College.

Fig. 300. Fin spines believed to have been from acanthodians similar to this one have been found in the Milwaukee Formation. After Nelson et al., 2016, p. 99.

Osteichthyes (Silurian–Holocene)

The Osteichthyes, traditionally known as "boney fish," is the most species-rich and diverse class of fishes, with members ranging from bass, to sea horses, to coelacanths, to lungfish, to (yes) people. As such, they now inhabit marine and freshwater environments, and virtually every terrestrial one. Again, this great diversity was not yet seen in the Devonian. Nelson et al. (2016) diagnosed the Osteichthyes as follows: "Skeleton, in part at least, with bone (endochondral or membrane bone); skull with sutures; lung(s) usually present, functioning either as air-breathing organs or as buoyancy-controlling swimbladders or both…" As discussed at the beginning of this chapter, this modern classification, now finding wide acceptance among taxonomists, expands the class to also include the tetrapods (four limbed vertebrates), which consist of the amphibians, reptiles, birds, and mammals. Two subclasses comprise the Osteichthyes: the Sarcopterygii (lobe-finned fishes and tetrapods) and the Actinopterygii (ray-finned fishes). All of the osteichthyan material from the Milwaukee Formation known to date are assigned to the Sarcopterygii.

Sarcopterygii (Silurian–Holocene)

This subclass has approximately 30,500 living species, of which only eight are fishes (Nelson et al., 2016). One of those fish is the famous "living fossil," the coelacanth. Teeth of the sarcopterygian fish *Onychodus* cf. *sigmoides* Eastman, 1907, with their distinctively curving, conoidal, pointed form, have been found in the Berthelet Member (Figs. 303–305). Cleland (1911) also listed an unnamed species of the sarcopterygian *Holoptychius* from the Berthelet Member; however, the fossils on which he based this identification cannot be located. The Milwaukee Formation included strata deposited contemporaneously with the appearance of tetrapods; however, unequivocal tetrapod fossils have not yet been recognized from the Milwaukee Formation.

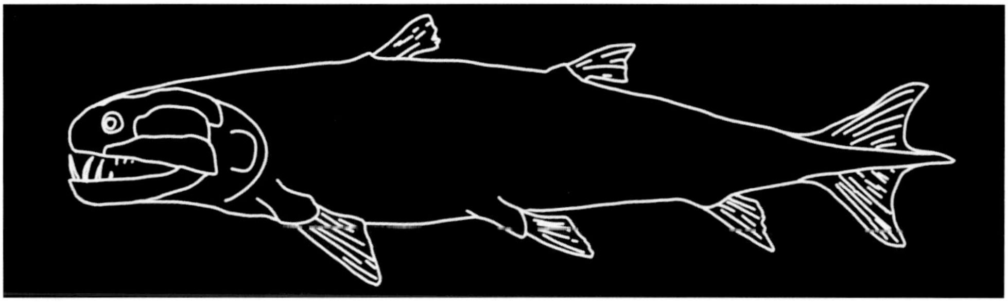

Fig. 303. The long, pointed teeth of the sarcopterygian (lobe-finned) fish *Onychodus* are occasionally found in the Milwaukee Formation.

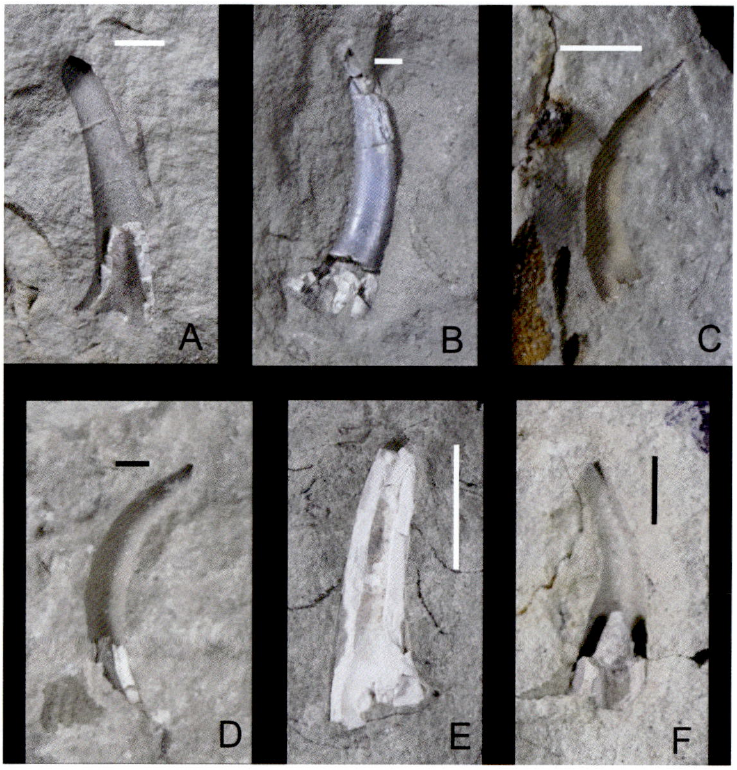

Fig. 304. Six teeth from *Onychodus* sp., Onychodontiformes: Onychodontidae; Berthelet Member. **A:** MPM P28175; collected by KCG; scale bar 2mm. **B:** MPM 15357; collected by GOR; scale bar 2mm. **C:** MPM P28176; collected by KCG; scale bar 5mm. **D:** MPM 747; collected by GOR; scale bar 2mm. **E:** MPM P28177; collected by KCG; scale bar 5mm. **F:** MPM P28178; collected by KCG; scale bar 2mm.

Fig. 305. Tooth of *Onychodus* sp., Onychodontiformes: Onychodontidae; specimen is longitudinally broken, revealing the pulp cavity and what may be carbonized pulp; MPM 477; Berthelet Member; figured in Cleland, 1911, Pl 47, Fig 5; collected by CEM; scale bar 2mm.

Chapter 21 — Fungi

At the present time, there is only one genus of the kingdom Fungi known from the Milwaukee Formation, though it was not classified as a fungus until relatively recently. Originally thought to be a conifer, and subsequently classified either as a green or brown alga or a lichen, *Prototaxites* (Silurian–Devonian) is, at the time of writing, considered to be a giant, terrestrial fungus (Hueber, 2001; Boyce et al., 2007). Reaching a width of 1.25m at its base and a height of 8.8m, *Prototaxites* was the tallest organism on land for the first fifty million years of its existence (Boyce et al., 2007).

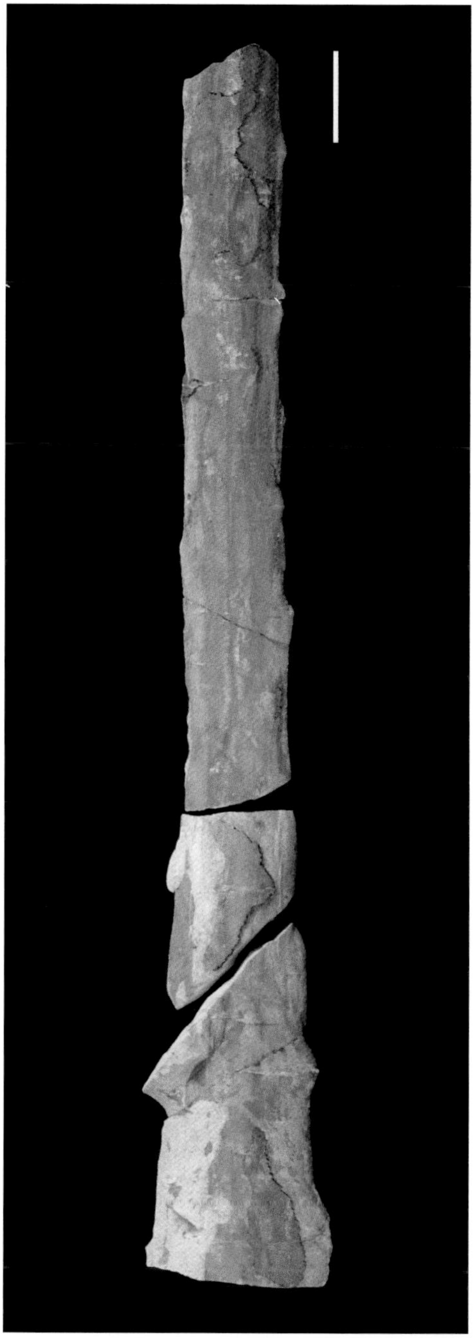

Fig. 306. *Prototaxites milwaukeensis* (Penhallow, 1908), Prototaxales: Prototaxaceae; MPM 402; holotype; Berthelet Member; figured in Penhallow, 1908, Pl 1 and Cleland, 1911, Pl 52, Fig 2; collected by CEM; scale bar 5cm.

In the Milwaukee Formation, this giant fungus is known from two fossils collected from the base of the Berthelet Member by Charles Monroe and originally given the name *Nematophycus milwaukeensis* Penhallow, 1908. It is now transferred to *Prototaxites*, being that *Nematophycus* was shown to be a synonym of that genus. Penhallow considered this specimen to be an alga, and indistinguishable from a species from the Hamilton of New York, now known as *P. crassus* (Penhallow, 1896), but he decided to provide it with its own name in spite of the potential synonymy.

The holotype is broken in three pieces and has a non-branching body approximately 68cm long when its three sections are reassembled and 5.5cm wide at its top; its base is 11.5cm wide (Figs. 306 & 307). Being broken at its top and bottom, the organism was undoubtedly longer (taller) before breakage, but was dwarfed by many other *Prototaxites* specimens alive during that time. It may have been a young individual. A second specimen (MPM 403), not examined by Penhallow but found by Monroe in the same beds as the holotype, appears to be a smaller section of another example of *P. milwaukeensis*, having a similar diameter, thickness, and thin exterior layer of coal.

Some land plants of the Late Middle Devonian, including some of those from the Milwaukee Formation, had surpassed even the tallest *Prototaxites* by this time, as will be discussed in the next chapter.

Fig. 307. *Prototaxites milwaukeensis* (Penhallow, 1908), Prototaxales; middle section of the holotype specimen shown on the previous page; MPM 402; scale bar 2cm.

Chapter 22 — Land Plants
(Plantae: Embryophytes)

Land plants (Embryophytes) (Ordovician–Holocene) were only rarely preserved in the Milwaukee Formation, yet were very distinctive and under-reported. Most, if not all of the known plant fossils from this formation appear to be fragments of tracheophytes (vascular plants) known as cladoxylopsids and lycopods. These plant groups can be arborescent (tree-like), as were those found in the Milwaukee formation, and are in fact believed to be the first trees and, up to that time, the tallest organisms on Earth, some of which towered over *Prototaxites*, the fungus that held that record in the Lower Devonian (Chapter 21).

Cladoxylopsids (Embryophytes: Tracheophyta: Cladoxylopsida) (Devonian–Carboniferous) are arboraceous tracheophytes typically having branches that dichotomize (branch) several times and a vascular system of radially directed plates. Cladoxylopsids include the famous giant stumps and trunks of the Middle Devonian Gilboa forests of New York. Cladoxylopsids appear related to the ferns and may be ancestral to them. The only published possible cladoxylopsid fossil from the Milwaukee Formation was thought to be a brown alga and thus named *Fucus bertheletensis* Penhallow, 1908. It consists of what Penhallow interpreted to be sixteen parallel straps. He stated, “In three separate places it may be seen that the members subdivide by a regular dichotomy, and the conclusion is justified that such dichotomy is characteristic of the plant as a whole.” This specimen is shown in Fig. 308. Penhallow was apparently not aware of the other bark fragments co-occurring with this specimen, many of which bear this same resemblance to parallel straps, and some having the “dichotomy” that he noted. He did not comment on the fact that the edges of similar “straps” are exposed on at least three 0.5–1cm layers below the surface he described and that the “straps” in all layers are oriented in the same direction (Fig. 308 C). That those lower (earlier) occurrences would all be oriented in the same direction is highly unlikely if representing straps of algae; however, this layered structure is typical of cladoxylopsid bark and it is found in many other co-occurring specimens, as illustrated in this chapter (Figs. 309–314). The specimen represented by *Fucus bertheletensis* is therefore herein regarded as a likely cladoxylopsid bark fragment. P. Kenrick (personal communication, 29-Apr-2018) suggested that it might be from a cladoxylopsid similar to *Eospermatopteris*.

Fig. 308. Fragment of cladoxylopsid bark, originally thought to be a marine alga and named *Fucus bertheletensis* Penhallow, 1908; holotype; Berthelet Member; collected by CEM; scale bar 2cm. **A:** MPM Pb399.1; surface of the slab also figured in Penhallow, 1908, Pl 2 and Cleland, 1911, Pl 53. **B:** MPM Pb399.2; counterpart of Pb399.1. **C:** Enlargement of region shown in white box in A, showing the fossil structure at several layers of the slab; white box is 8.5cm wide.

Fig. 309. Cladoxylopsid? bark; UWM-GM 9757; Berthelet Member; scale bar 1cm.

Fig. 310. Cladoxylopsid? bark; UWM-GM 9773; Berthelet Member; scale bar 1cm.

Fig. 311. Cladoxylopsid? bark; UWM-GM 25826; Berthelet Member; scale bar 1cm.

Fig. 312. This small piece of cladoxylopsid? bark is one of the few such fossils collected from the Lindwurm Member. WESM-KCG 1114; collected by KCG; scale bar 5mm.

Fig. 313. Large piece of cladoxylopsid? bark indicating a trunk diameter of at least 28cm; UWM-GM 25808; Berthelet Member; enlargement of the region within the white rectangle is shown below. It should be noted that the parallel structures of this specimen are narrower than those of the other cladoxylopsid? specimens shown in this book.

Fig. 314. Close-up of the region outlined by the white rectangle shown on the preceding image of cladoxylopsid? bark; scale bar 1cm.

Lycopods (Embryophytes: Tracheophyta: Lycophytina) (Silurian–Holocene) are tracheophytes having leaves arranged in a dense, regular spiral pattern that results in distinctive scar patterns in the deciduous members of the group; leaves are narrow and typically with one unbranched vein (Gensel & Andrews, 1984). While some are herbaceous or shrubby in form, many living in the Middle and Late Devonian (such as those found in the Milwaukee Formation) and Carboniferous were woody and arborescent, with some attaining heights of up to 35m. The Middle Devonian occurrences, such as that from the Milwaukee Formation, together with the co-occurring possible cladoxylopsids discussed in the previous chapter, thus mark the planet's first appearance of trees and forests.

The Greene collection houses several carbonized lycopods from the lower region of the Berthelet Member (Monroe & Teller, 1899) considered to be arborescent lycopod fragments, including bark, axes (branches), and leaves (Figs. 315–319). The surface of one piece of bark clearly shows the typical lycopod leaf scars (Fig. 320).

Fig. 315. Unusually well-preserved axis (branch) of a lycopod, showing leaves along edges and leaf scars; UWM-GM 9768; Berthelet Member; scale bar 1cm.

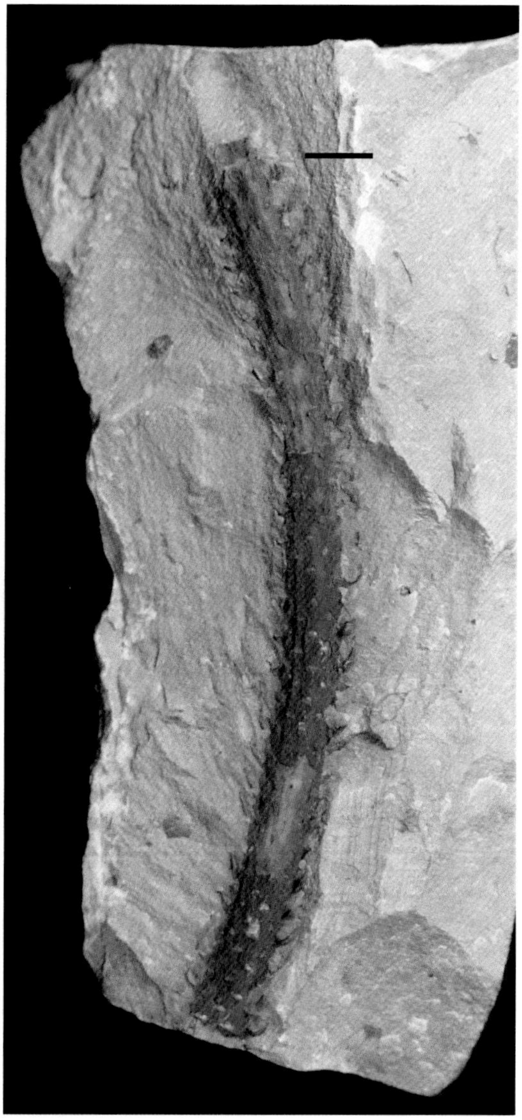

Fig. 316. Axis (branch) of a lycopod, showing leaves along edges and leaf scars; UWM-GM 9761; Berthelet Member; scale bar 1cm.

Fig. 317. Axis (branch) of a lycopod, showing leaves along edges and leaf scars; UWM-GM 9774; Berthelet Member; scale bar 1cm.

Fig. 318. Axis (branch) of a lycopod; UWM-GM 9771; Berthelet Member; scale bar 5mm.

Fig. 319. Unidentified plant material, possibly lycopod leaves; UWM-GM 9760; Berthelet Member; scale bar 1cm.

Although a few specimens of these plant fossils are quite well preserved, the great variety of details due to preservational factors makes precise identification of these fragments difficult or impossible. In the case of lycopod occurrences, some authors avoided specific identification and instead refered to these fragmentary occurrences simply as "lycopod fragments" (Grierson & Banks, 1963; Gensel & Andrews, 1984). Accordingly, we have chosen not to attempt specific identification of these plant remains, and encourage further investigation from plant specialists.

Roberts (1972) reported the occurrence of sporangia (structures within which spores are produced) from the Milwaukee Formation. Whether they were from fungi or land plants was not noted, and nothing else has been published on this occurrence. These sporangia cannot be located.

Except for the above-mentioned studies of Penhallow, Cleland and Roberts, only a few other works have been published that mentioned the land plants or sporangia of the Milwaukee Formation (e.g., Chamberlain, 1883; Monroe & Teller, 1899; Kluessendorf, Mikulic & Carmen, 1988). As far as the authors can determine, Cleland's publication was the last to include illustrations of the plant fossils, and those were only duplications of Penhallow's (1908) images.

Fig. 320. Three views of a piece of lycopod bark; UWM-GM 51372; Berthelet Member; scale bar 1cm. **A:** "front" side, showing the leaf scars typical of lycopod bark. **B:** "back" side. **C:** side view, showing the layered appearance typical of the other tree bark from the Milwaukee Formation.

Chapter 23 — Trace Fossils

Trace fossils are the fossilized markings or structures made by the activity of an organism on or within a substrate or another organism. They are common in all members of the Milwaukee Formation.

Palaeophycus (Fig. 321) is the most common ichnogenus in the Lindwurm Member. In the Milwaukee Formation, these are 0.5–1cm wide, sub-horizontal burrows that are filled with sediment that is of the same composition as the surrounding matrix, but often of a darker shade. They sometimes contain narrow bands of a still darker shade along their lengths. *Palaeophycus* is generally attributed to a feeding burrow from a worm-like animal, such as a polychaete. Pemberton & Frey (1982) suggested their maker to be a predaceous or suspension-feeding animal. As discussed earlier in this book, scolecodonts found in the Milwaukee Formation indicate the presence of annelid worms, and may suggest the producer of these traces.

As seen in the images in Chapter 14, the majority of the cephalopods in the Milwaukee Formation have curving trough-like structures across their surfaces that appear to be burrows (Fig. 322). These structures appear to have been produced by the burrowing activity of some type of scavenging worm after burial of the shell (Hilderbrand & Zambito, 2015).

Vermiformichnus represents infillings of shell borings, possibly from parasitic spionid polychaete worms. One peculiar specimen in the Milwaukee Public Museum (Fig. 323) that may belong to this ichnogenus is labeled "Vermes." Borings in the form of drill holes through the shells are also found on rare occasions in brachiopods and bivalves of this formation (Fig. 175).

Zoophycos (Ordovician–Holocene) is a more-or-less spiral vertical burrow in marine sediment produced by a moving or feeding (sediment mining) organism. They were probably produced by a polychaete or non-segmented worm (Miller et al., 2009), although other organisms such as sea pens have also been suggested for much later and recent occurrences (Bradley, 1973). Certain burrows found in the North Point Member have been identified as *Zoophycos* (Mayer, 2001).

Arthrophycus? (Cambrian–Carboniferous) is a finger-sized, typically branching burrow. Many examples of this ichnogenus (such as specimens found in the Berthelet Member, including that shown in Fig. 324) were commonly identified as "fucoids" due to their resemblance to the brown alga *Fucus*. The Milwaukee fossils lack the segmentation typically found on *Arthrophycus*. Their makers are unknown.

Chondrites (Cambrian–Holocene) is a narrow, branching burrow in marine sediment. Although they are still produced in deep water sediments, their producers have escaped detection. As with *Phycodes*, *Chondrites* has been found in the Berthelet Member (Fig. 325).

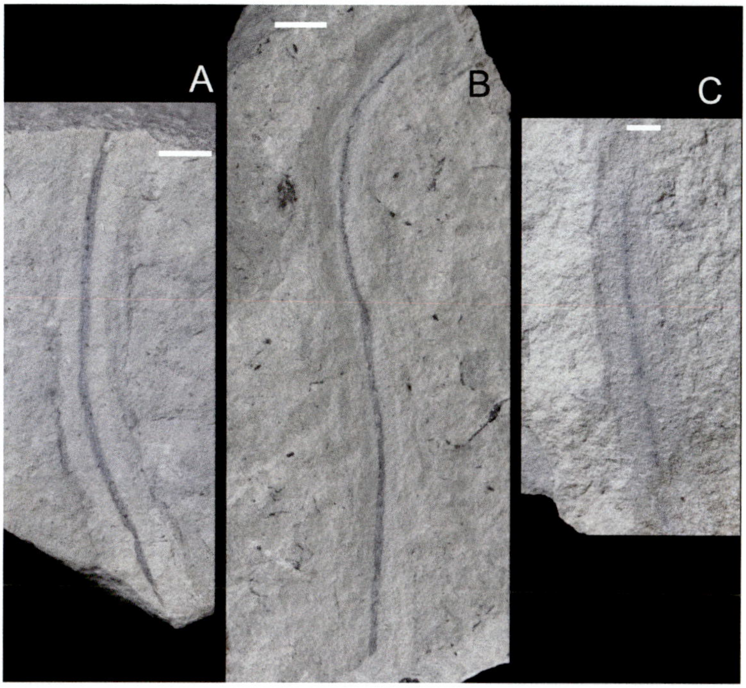

Fig. 321. Three apparent trace fossils identified as *Palaeophycus* isp.; Lindwurm Member. **A:** KCG collection, uncataloged; collected by KST; scale bar 5mm. **B:** MPM P28183; collected by KCG; scale bar 5mm. **C:** MPM P28184; collected by KCG; scale bar 2mm.

Fig. 322. Cephalopod with numerous burrows on its surface; UWM-GM uncatalogued; Berthelet Member; scale bar 1cm.

Fig. 323. Apparent fillings of worm borings (*Vermiformichnus?* isp.) associated with a brachiopod; MPM 15149; Lindwurm Member; collected by GOR; scale bar 5mm.

Fig. 324. This fossil is labeled "fucoids," but it is here interpreted to be the burrow *Arthrphycus?* isp.; Berthelet Member; MPM P26723; scale bar 1cm.

Fig. 325. *Chondrites?* isp.; UWM-GM 9770; Berthelet Member; scale bar 5mm. This trace is flat and branches in only one plane, in contrast to the trace shown in the previous image.

Chapter 24 — Unidentified Body Fossils

Among most fossil biota are body fossils that are not readily identifiable. Some may simply be fragments of an organism that lack diagnostic features, while others represent portions of known taxa not recognized due to unusual states of preservation, and in still other cases they may be from taxa not previously known from that biota. Such fossils are common in the Milwaukee Formation—especially the Lindwurm Member. Whether representing remains of plants, arthropods, fish, bryozoans, hydrozoans, or organic matter of other origins, a selection of unidentified body fossils is shown here (Figs. 326 & 327).

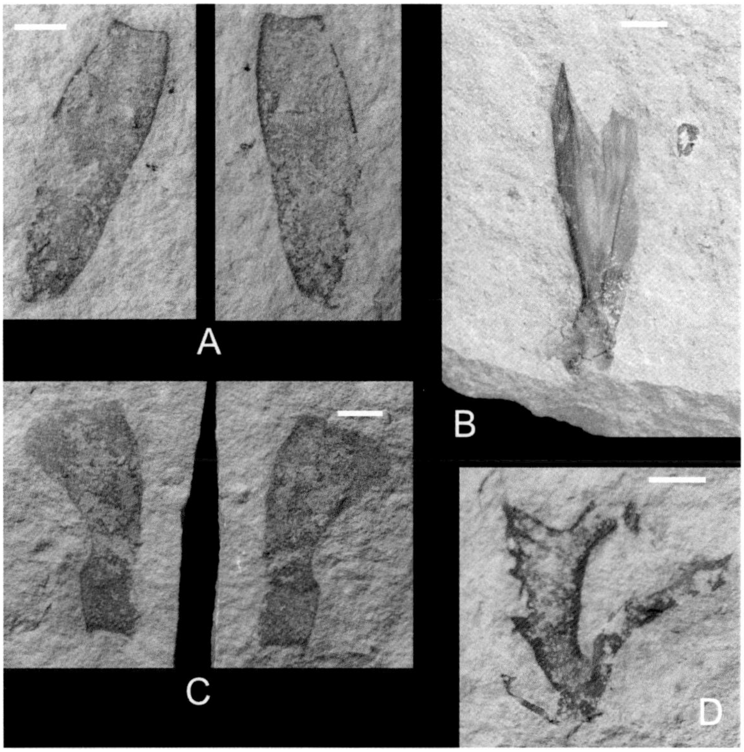

Fig. 326. Unidentified body fossils; Lindwurm Member. **A:** MPM P28185, part and counterpart; collected by KCG; scale bar 5mm. **B:** MPM P28186; collected by KCG; scale bar 2mm. **C:** MPM P28187, part and counterpart; collected by KCG; scale bar 5mm. **D:** MPM P28188; collected by KCG; scale bar 5mm.

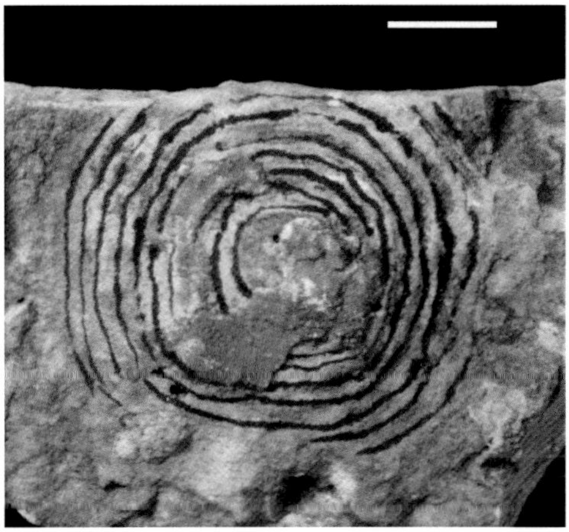

Fig. 327. The pattern of concentric cirles on this unusual fossil appears similar to growth lines, such as those of certain lingulid brachiopods and the bryozoan *Lichenalia*; however, in neither case have the growth lines been observed to be carbonized, as these appear to be. MPM P28194; Berthelet Member; scale bar 1cm.

Chapter 25 — Fossil Collecting in the Milwaukee Formation

Nearly all of the Milwaukee Formation fossils preserved in museums were collected from a very small number of sites in Milwaukee County and the heyday of this collecting has long past. The Lake Michigan shore outcrops at Whitefish Bay were undoubtedly the first discovered and beginning with Lapham, have been visited by nearly every geologist to study the Milwaukee area Devonian rocks over the last 175 years. Although most of the more common Milwaukee Formation species have been found there, including some not reported from other sites, few specimens from this locality can be identified in existing collections. Currently located on private property and being very difficult to get to this site, it is not likely to be a good future source of specimens. Further complicating collecting along the shore in recent decades, shore line protection structures have covered some of the few remaining outcrops.

The vast majority of Milwaukee Formation fossils were collected along the Milwaukee River exposures. From their opening in 1876 to their closing in 1911, the quarries of the Milwaukee Cement Company produced the vast majority of fossils collected from the Berthelet Member (zones A and B). Natural outcrops of these rocks existed there before the quarries were started, and by 1844 Lapham had already assembled a good collection of the more common Milwaukee formation fossils from them. After the company ceased operations, the extensive remaining quarry walls and related rock exposures provided excellent opportunities for collecting the upper Berthelet and lower Lindwurm members by both scientists and the general public. This situation has slowly declined and, after over one hundred years of landscaping in Estabrook Park and garbage dumping in the Cement lake area, almost all exposures of the Milwaukee Formation have now been destroyed (Chapter 28). Many of the remaining rock exposures are not even natural outcrops. The best example of this is the "falls" area of Estabrook Park, which, along with being the type locality of the Milwaukee Formation, is the remains of an old quarry wall. Unfortunately, little planning has been undertaken to preserve any of the once extensive collecting sites for scientific research or public education.

Ironically and unfortunately in some respects, the most tangible preservation action that has been taken is also that which makes further fossil collecting, fieldwork and resulting research impossible. Estabrook (Fig. 328) and Lincoln Parks are, at the time of writing, off-limits to fossil, rock, or mineral collecting, on the basis of being non-renewable resources, in accordance with Milwaukee County Ordinances. No exceptions for research, education or any other purpose are made (B. Russart, Natural Areas Coordinator, Milwaukee County Parks, personal communication, 28-Jun-2018). Heavy fines are imposed for infractions.

However, that doesn't mean that it is impossible to collect Milwaukee Formation fossils. The rocks of this formation are occasionally excavated in construction projects providing some possibilities. The best example was during the construction of the Milwaukee Metropolitan Sewerage District's Deep Tunnel project of the 1980's and 1990's. This project provided the most important opportunity of the century to study the geology of the Devonian and Silurian rocks of Milwaukee County. Although it was not possible for public collecting at these construction sites, some of the locations where excavation material was being dumped could be visited with permission of the land owners. Other more widespread opportunities to collect Milwaukee Formation fossils can be found.

The glacial sediments of the region contain varying amounts of Devonian rocks that have produced excellent fossils. This has been known since the mid 1800's and as an example the type specimens of the crinoid *Melocrinites nodosus* described by James Hall in 1861 had been collected from glacial sediments in Milwaukee County by I.A. Lapham. Throughout much of the area, south-westward moving Quaternary ice sheets transported Devonian rocks from Lake Michigan and Milwaukee County to surrounding areas in southeastern Wisconsin and northeastern Illinois. Small weathered out specimens of Lindwurm specimens can be very common, especially crinoid ossicles probably from *Melocrinities*, along with bryozoans and brachiopods similar to those in figures 20 and 21. Rare small fish parts have also been found.

Fig. 328. The type locality of the Milwaukee Formation at Estabrook Park, Milwaukee, Wisconsin.

Milwaukee Formation fossils preserved in glacial deposits have been found in many locations.

- G.O. Raasch in unpublished 1928 field notes recognized Milwaukee Formation fossils in the vicinity of Port Washington in Ozaukee County.
- KCG collected *Eldredgeops*?, brachiopods, bryozoans, crinoid ossicles, and tentaculitids from the Lindwurm Member in a sand and gravel pit and housing excavations in Brookfield, Waukesha County (43° 3' 38.042" N 88 N, 88° 6' 23.323" W) in the late 1950s and 1960s.
- DGM collected Lindwurm fossils during the late 1950s and early 1960s at Tess Corners (center section 2, T5N, R20E), Little Muskego Lake, (northwest ¼ section 4 T5N, R20E, and in Muskego (northwest Section 5, T5N, R20E), all in Waukesha County.
- JK and DGM collected specimens of *Melocrinites* in northwestern Milwaukee County during the early 1970s, (southwest sec. 30, T8N, R21E).
- Curt Klug collected specimens near Wind Lake in Racine County and other Lindwurm specimens were observed along the Root River near Racine (Center section 6, T3N,R23E).
- Milwaukee Formation fossils have also been found in northern Lake County Illinois along the Lake Michigan shore.

These are locations where Milwaukee Formation fossils have been found in glacial deposits over the years through random searches. There are sure to be numerous other locations in southeastern Wisconsin and northeastern Illinois, especially along the shore of Lake Michigan. Any weathering exposures of glacial sediments may contain small fossils of the Lindwurm and North Point Members or larger boulders of the Berthelet.

Permission must be granted by the property owners before attempting to collect. Due to the scarcity of collecting sites, if you are lucky enough to obtain permission, conservative methods should be used. When Gass and Mikulic were still collecting in Estabrook and Lincoln Parks, the only tools used were fingers and a whisk broom.

Chapter 26 — Preparation Methods

Preparation of fossils (i.e., removal of the matrix that surrounds the fossil) has included a number of methods influenced by the type of preservation involved, the state of technology at a particular point in time, and how much effort and money the preparator is willing to spend. Modern fossil preparation techniques, such as air abrasion, were not yet available when the majority of the Milwaukee Formation fossils were collected. Instead, most of the fossils from the shaley layers of the Lindwurm Member were prepared by scraping away most of the matrix with a dissecting needle or similar instrument, followed by scrubbing the fossil for minutes or hours with a moist toothbrush, and rinsing the fossil under running water. Many times, especially during the early 1900s, the fossil was first soaked or boiled in a solution of potassium hydroxide (caustic potash) (Bather, 1908; Cleland, 1911). The downside of using KOH is that if it is not adequately rinsed following preparation, it attacks the matrix and fossil, destroying the fine details, leaving a distracting white residue, and sometimes disintegrating the entire fossil (Fig. 329). It should also be noted that if this method is to be considered, proper personal protection equipment and ventilation must be used. This is not to suggest that the more conventional methods are any less likely to damage the fossil or harm the preparator. Careless use of dissecting needles has scratched many fossils and stuck many hands. Likewise, the hammer has smashed its share of fossils and fingers, as the authors can attest.

Fig. 329. The effects of careless use of caustic potash (KOH) on trilobite fossils. These two trilobites were apparently soaked in a KOH solution and inadequately rinsed. After several years of storage, not only had the matrix crumbled away, but, in the case of the specimen on the right, the fossil itself had almost completely disintegrated; MPM 25378 (specimen on left); Lindwurm Member; collected by MOE.

Possibly the least injurious and most effective preparatory method for calcareous fossils in the shaley layers of the Lindwurm Member is air abrasion. After removing the bulk of the matrix with hand power tools or hammer and chisel, abrasive particles that are of the appropriate hardness and size are used to sand-blast away the remaining matrix, thus revealing fine details without breaking, abrading or otherwise damaging the fossil (Fig. 330). Use of this method is limited due to the cost of the necessary equipment as well as the level of skill required.

Most of the methods that are appropriate for preparing fossils from the Lindwurm Member may be inappropriate for the Berthelet Member, being that in the latter case the fossils are generally in the form of internal and external molds in hard, dolomitic limestone. The use of dissecting needles, air abrasion, and KOH generally rely on the fossil being harder or more chemically resistant to those methods than the matrix is. Because internal and external molds are often of equally informative value and are both composed of the same material, preparators avoid those methods when dealing with fossils preserved as molds. Due to the gap present between the molds, the hammer-and-chisel method and use of hand power tools are the most preferred and effective methods of preparing fossils preserved in this way.

Fig. 330. Preparation stages for a *Melocrinites nodosus spinosus* calyx from the Lindwurm Member; KCG collection uncatalogued; collected by KCG; scale bar 5mm. **A:** As found, the outline of the calyx and indications of the plate spines can be seen embedded in the soft shaley matrix. **B:** Fossil is largely uncovered using a power hand tool for gross matrix followed by an air abrasor for the finer details; photograph courtesy of David Comfort. **C:** The completely prepared specimen; prepared by David Comfort.

None of the above methods pertain to microfossils. Instead, for microfossils of a different chemical composition from the matrix, acids or other chemicals are used. The conodonts shown in Figs. 251 & 252, being phosphatic in composition, were prepared by dissolving crushed samples in a 10–15% acetic acid bath for a week or so, and washing the resulting sludge through sieves (D. Schumacher, personal communication, 11-Nov-2017). For the foraminifers shown in Fig. 23, Roberts (1972, p. 40) used acetic acid, hydrochloric acid or hydrogen peroxide, depending on the sample. Again, if these methods are to be used, adequate ventilation and appropriate caution and personal protection equipment should be employed.

Chapter 27 — Major Collections

Long before fossil collecting in the Milwaukee Formation became nearly impossible due to imposed laws and inaccessible or depleted sources, several individuals were fortunately and diligently collecting fossils on an enormous scale. These were engineer and naturalist Increase A. Lapham, pharmacist Thomas A. Greene, attorney Charles E. Monroe, and businessman Edgar E. Teller (Mikulic, 1983) (see Chapter 1). Had these "Gentlemen Paleontologists" not assembled these collections before the virtual disappearance of the Milwaukee Formation, our knowledge of its biota would be rudimentary at best, and hopes of furthering our understanding would be severely reduced. Some of this material has been lost, while large parts of these collections in most institutions have not been cataloged, even though they have been held for over 100 years. Following are descriptions and locations of these collections.

Increase A. Lapham

Lapham's collection was important for including perhaps the first fossil collection assembled in Wisconsin and for signaling the presence of Devonian rocks in Wisconsin. He was the first person to study the bedrock geology and paleontology of the Milwaukee area and was instrumental in bringing these rocks to the attention of James Hall and other eastern geologists. Lapham was very generous with his specimens and gave material to a number of scientists around the country. He also did some trading, and a few of his fossils with their distinctive labels can be found in many museum collections to this day. After his death in 1875, most of his collection consisting of 10,000 fossils, minerals, shells, meteorites, and Native American artefacts, was purchased by The University of Wisconsin; however, it was destroyed in a fire in 1884 (Mikulic, 1983). The Milwaukee Public Museum was given a few hundred specimens that were originally retained by his family after his death. This material contains a few Milwaukee Formation specimens.

Thomas A. Greene

Greene's collection has the distinction of being the sole major nineteenth century Milwaukee area fossil collection to remain intact and in the area. The Thomas A. Greene Memorial Museum (aka Greene Museum) was built at what is now the University of Wisconsin-Milwaukee to house this collection of around 75,000 fossils, minerals and other materials, many of which are from the Milwaukee Formation (Mikulic, 1983). Because of their uniqueness and importance, in 1993 the Greene Museum and its collections were together designated a National Historic Landmark in the History of Science (Kluessendorf & Mikulic, 1993a; Mikulic, 1991; National Park Service, 2016). Unfortunately, the collection was moved out of the museum, which has jeopardized its continuing Landmark status, and is now maintained at Lapham Hall in a compactor storage system (Fig. 331) and on display in a small museum area. Plans are being made to transfer the data from the paper catalogs to a searchable digital database (S. Dornbos, personal communication to K.C. Gass, 12-Sep-2017).

Charles E. Monroe

In addition to being a diligent collector of Devonian fossils from the Milwaukee area, Monroe also published some of the early papers on Devonian fossils and stratigraphy of that area, and he served as honorary curator of paleontology at the Milwaukee Public Museum from 1897 to about 1922 (Mikulic, 1983). His hand-written entries can be seen in the fossil catalogs from that time. He donated most of his extensive collection to MPM, where it still resides (Fig. 332). At the time of writing this book, data were currently being transferred from MPM's paper catalogs to a searchable digital database (P. Currough-Burke, personal communication to K.C. Gass, 10-Jul-2017).

Edgar E. Teller

In addition to amassing a large collection of fossils, Teller also published early papers on the geology and paleontology of the Wisconsin Devonian, one of which was co-authored with Monroe. He concentrated most of his collecting efforts on the Devonian fossils found in the Milwaukee Cement Company quarries and in Silurian localities in the region. Teller gave a large part of his massive collection to the Walker Museum

Fig. 331. Greene Geological Museum and fossil storage at the University of Wisconsin-Milwaukee. **A:** The original Thomas A. Greene Memorial Museum building as it looked in July, 2018. This building is now part of UWM's Sam and Helen Stahl Center for Jewish Studies. **B:** Compactor storage system in Lapham Hall that currently houses the Greene collections.

Fig. 332. Main storage area for Devonian fossils at the Milwaukee Public Museum.

at the University of Chicago (now part of the Field Museum). Other parts of his collection can be found in the Royal Ontario Museum in Toronto, and the Buffalo Museum of Science. Teller's remaining 100,000 specimens, including all of his type specimens, were donated to the National Museum of Natural History after he died, where they still reside (Mikulic, 1983). This collection cannot be examined intact, having been split between several departments and buildings. Much of it has been cataloged in a searchable on-line database; however, some of it has never been cataloged at all.

Gilbert O. Raasch

Raasch is the most important of the twentieth century collectors of Milwaukee Formation fossils. He did extensive research on the geology of the Wisconsin Devonian. His collections (MPM) of the North Point and Lindwurm fossils are among the largest known. Although he never finished this work, he was long regarded as one of the experts on the Devonian rocks of North America (Mikulic & Kluessendorf, 2001).

Other collections of significance are those assembled by Fisk Holbrook Day (now residing in MCZ), Herdman F. Cleland (now at Williams College), and Joseph G. Emielity (MPM). Many other individuals, including, but not limited to, those referenced in the Preface, also have material in the aforementioned museums, the University of Wisconsin Geology Museum (Madison), the Weis Earth Science Museum, and other public and private repositories.

Most of these collections have not undergone a thorough modern study, and much of the Greene collection has never been studied at all. The authors therefore hope that readers of this book will fully utilize this material in continuation of the much-needed research of the diverse Milwaukee Formation biota.

Chapter 28 — The Cement that Made Milwaukee Famous

Donald G. Mikulic and Joanne Kluessendorf
Weis Earth Science Museum
University of Wisconsin-Oshkosh, Fox Valley Campus
Menasha, Wisconsin

Milwaukee, Wisconsin is well known historically as a center of American industry and commerce. During the late 19th and early 20th centuries, it was the home of some of the more prominent manufactories and breweries in the country. Largely forgotten however is its role as one of the leading late 19th century cement producers in the United States (Fig. 333). From 1876 until 1911 the Milwaukee Cement Company dominated the Midwestern cement market through the manufacture of Milwaukee Hydraulic Cement, then considered to be the best natural cement in the country and recognized as the standard for use in federal government and other municipal projects (Kluessendorf & Mikulic, 2008). From large modern plants along the Milwaukee River, Milwaukee Hydraulic Cement was shipped over a vast region extending more than 1200 miles from "Detroit and Cleveland on the East to Kansas City and Denver on the West" (Greene & Berthelet, 1949) and from Canada on the north to Kentucky on the south. Although this enterprise had considerable economic importance in Wisconsin, it also contributed to important scientific and educational interests. Primarily through the excavation of cement rock at the Milwaukee Cement Company's quarries and mines, a unique opportunity existed to make large fossil collections and conduct scientific research on the Middle Devonian rocks of the Milwaukee Formation. As a result of the large-scale quarrying along with the rarity of Devonian rocks in the region, Milwaukee was at the time one of the most important sites for scientific studies on the Devonian in the Midwest. Without the activities of this industry, much less would be known about the geology and paleontology of the Milwaukee Formation, and this book would probably not have been written. This chapter discusses how the region's geology and settlement, combined with changing technology and economic conditions, first created and then destroyed a major industry.

Stone for construction

During the early 19th century, wood was relatively well distributed and clay used in brick making could be found in many communities, but sources of stone needed to manufacture critical construction products such as cement, lime, and building stone were for geologic reasons few and far between. In addition, with the availability of only simple excavation equipment, it was difficult to produce stone resources on a large scale, and the poorly developed transportation systems of the time made shipping of these heavy and bulky commodities very difficult, expensive, and in some cases impossible. As a result, during much of the 1800s, only local sources of these critical stone based materials were available throughout much of the Midwest and when found, they were used regardless of their quality.

Early 19th century Milwaukee is a typical example of the geologically restricted distribution of these types of mineral resources. As a place where the bedrock surface is generally buried by thick deposits of unconsolidated Quaternary sediments, it was difficult for most early settlers to find nearby rock exposures where they could quarry or purchase stone construction materials. Building stone and lime were commercially produced in the region but only at a limited number of geologically controlled and irregularly distributed locations. The poorly developed roads of the time along with the lack of railroads in the area presented problems in moving these materials even over short distances. While lime and building stone were usually available at many sites in southeastern Wisconsin, initially there was no cement manufactured from local resources. Without cement, lime mortar was the primary binding material used in construction. If a person wanted to build a brick or stone building, lime mortar was a more economical and widely distributed alternative to importing cement from great distances. It was relatively easily made requiring less experience and equipment to manufacture. For example a farmer with access to almost any kind of limestone could with limited experience and investment build his own kiln and using local wood burn all

334 MILWAUKEE'S GREAT INDUSTRIES.

DAILY CAPACITY, 4,000 BARRELS.

OFFICE OF

MILWAUKEE CEMENT COMPANY.

OFFICERS.

H. BERTHELET, President.
T. A. GREENE, Vice-President.
JNO. JOHNSTON, Treasurer.
GEO. S. BARTLETT, Secretary.
JOS. R. BERTHELET, Superintendent.

BOARD OF DIRECTORS

JOHN JOHNSTON.
SML. MARSHALL.
T. A. GREENE.
HENRY BERTHELET.
JOS. R. BERTHELET.
WM. PLANKINTON.
B. K. MILLER.

SOLE MANUFACTURERS OF THE CELEBRATED

Milwaukee Hydraulic Cement

Office: Room 7, Plankinton Bank Building,

MILWAUKEE, WIS.

Milwaukee Cement is now conceded to be the best natural Hydraulic Cement in the American market. In uniformity of manufacture, reliability and ultimate strength, there is no superior cement made from natural rock, in this country or Europe.

We would call your attention to the bird's-eye view of the Company's works and property. Our works are the largest, most modern and complete for the perfect manufacture of HYDRAULIC CEMENT of this class in any part of the world. Accurate, impartial, scientific and practical tests, many thousand in number, made by and under the supervision of our Superintendent of Manufacture, J. R. Berthelet, an educated engineer and a gentleman of fifteen years' experience in the manufacture and practical manipulation of Cement, demonstrate that MILWAUKEE CEMENT is now the Standard Cement of the United States for all private and public work requiring a high grade of mortar and concrete.

We extend a cordial invitation to all Engineers, Architects, Contractors, and others interested to visit our extensive works, and satisfy themselves as to our method of manufacture, and they will be convinced that this Company manufacture as uniform and thoroughly reliable a product as modern skill, ingenuity and business capacity can produce.

With superior facilities for the prompt, safe and cheap shipment of our celebrated brand of HYDRAULIC CEMENT, we are able to meet every demand upon us on your part promptly, and, we believe, to the satisfaction of all.

Contractors to the United States Government for furnishing 100,000 barrels of Milwaukee Cement used in their extensive construction at Sault Ste. Marie.

MILWAUKEE CEMENT CO.

Fig. 333. Advertisement for the Milwaukee Cement Company from Anderson & Bleyer, 1892.

of the lime he needed for mortar. It was even possible to burn limestone glacial boulders in a pit dug into the ground to make lime. Manufacturing cement, however, requires rocks with a very specific composition which usually have a very geologically limited distribution. In addition, cement can only be manufactured in more complex and expensive facilities which requires a crew of knowledgeable individuals to operate and generally used coal as a fuel. For these reasons lime kilns were a common feature throughout much of the 19th century Midwest, while cement plants were very rare. As larger and more durable buildings along with other structures were being built during the late 1800s, lime mortar became less suitable as a binder and, as a better replacement, the need for cement in large scale construction projects greatly increased. As a result, the geologic controls on the availability of the stone resources needed to manufacture cement became more important. The closest cement producing locations to Milwaukee were at Utica, Illinois and Louisville, Kentucky. However, with the lack of a good land-based transportation system during the first half of the 19th century, cement, even in small amounts, could not be shipped cheaply over land to Milwaukee from either place. Instead limited and still expensive supplies of cement were imported by boat from more distant eastern locations on the Great Lakes such as New York. Until railroad transportation became available, true cement was probably a rarely used building material in Milwaukee.

The discovery of cement rock was economically important for Milwaukee area construction as local cement production greatly lowered the shipping cost of this critical building material, while it also became the basis of an important export commodity. Geology played a critical role in how Milwaukee's cement rock was discovered and how the industry developed. The "limestone" (actually dolomite or dolostone, which is a magnesium rich carbonate rock) of the Middle Devonian Milwaukee Formation, only occurs in a small area of northern Milwaukee County (Fig. 334), and continues east beneath Lake Michigan (see Chapter 2 for details on the geology of the Milwaukee Formation). As true throughout most of eastern Wisconsin, the bedrock surface in Milwaukee County is generally buried beneath a thick blanket of unconsolidated Quaternary glacial or lake sediments. Natural outcrops of the Devonian age rocks in this area were therefore rare and small in size but had been known for centuries. As early as the 1670s, French visitors to the region reported finding scattered rock exposures containing "pitch" (natural asphalt) along the Lake Michigan shore near present day Milwaukee. They used this material to caulk their canoes. It is possible that they were referring to an exposure of the Milwaukee Formation which contains asphalt, and can still be seen at Whitefish Bay (Thwaites, 1900). No other rock outcrops containing asphalt are known along the Lake Michigan shore of Wisconsin. In the 1830s, at the start of local European settlement, outcrops of the same rocks were also found extending for almost a mile along the Milwaukee River between what is now Capitol Drive and Hampton Avenue. Here much of the river flowed directly on the bedrock surface with a few low outcrops occurring along the river banks. Initially a few quarries were opened in this area to produce lime or building stone, but the clay content of the cement rock was too high to make good lime and it was too thin bedded and soft to produce good building stone. Although inferior in the quality needed to make these common stone products, it was found later that these rocks had excellent properties needed to make natural cement. In geologic literature and fossil collection catalogues, the outcrops and quarries in this area are referred to by a variety of names including: Humboldt, Lindwurm, Berthelet, Cementville, Milwaukee River, and Milwaukee.

Natural cement

Milwaukee Hydraulic Cement is a natural cement which differs in several important ways from the Portland (artificial) cement that is now universally used in construction. Natural cement is made by calcining (burning at a high temperature) argillaceous (having up to 25% clay minerals) carbonate rocks such as limestone ($CaCO_3$) or dolomite in a kiln which changes its chemical composition by driving off carbon dioxide. The resulting clinker (CaO) is then ground into a fine powder making natural cement. Adding water to the cement will cause it to harden through reversing the chemical process and when mixed with aggregates such as crushed rock will make concrete. During much of the 19th century, natural cement rock was the source of most of the cement used for heavy construction in the United States. While many carbonates are argillaceous, only a few have the correct carbonate to clay ratio and mineralogy needed to produce

Fig. 334. Areal geology of a portion of northeastern Milwaukee County. Purple striped areas (labeled Dm) portray locations along the Milwaukee River and the Lake Michigan shore at Whitefish Bay where the Milwaukee Formation occurs at or near the ground surface. From Alden, 1906.

a good cement. The argillaceous component of natural cement gave it an important characteristic as a hydraulic (water lime) cement and it will set under water. This is in contrast with non-hydraulic cements such as the commonly used lime mortar which was a cheaper, more easily made, and a widely available binding material. Although natural cement rock was the primary cement of the 19th century, during the early part of the century experimentation in Europe had led to the development of Portland (artificial) cement in which a clay free limestone/dolomite is artificially combined with clay or shale before calcining. This process produces a higher quality hydraulic cement compared to that made from natural cement as many of the former's characteristics can be altered and controlled while those of the later cannot. Although it had superior qualities, the use of Portland cement only slowly replaced the use of natural cement towards the end of the 1800s. In general, Portland cement had higher manufacturing and shipping costs which limited its use, and there were few places where, for geological reasons, the correct types of limestones and shale could be found in close proximity to a cement plant. It was only with the use of rotary kilns, which provided a more uniform and efficient way of burning of the rock, along with improved transportation systems, that

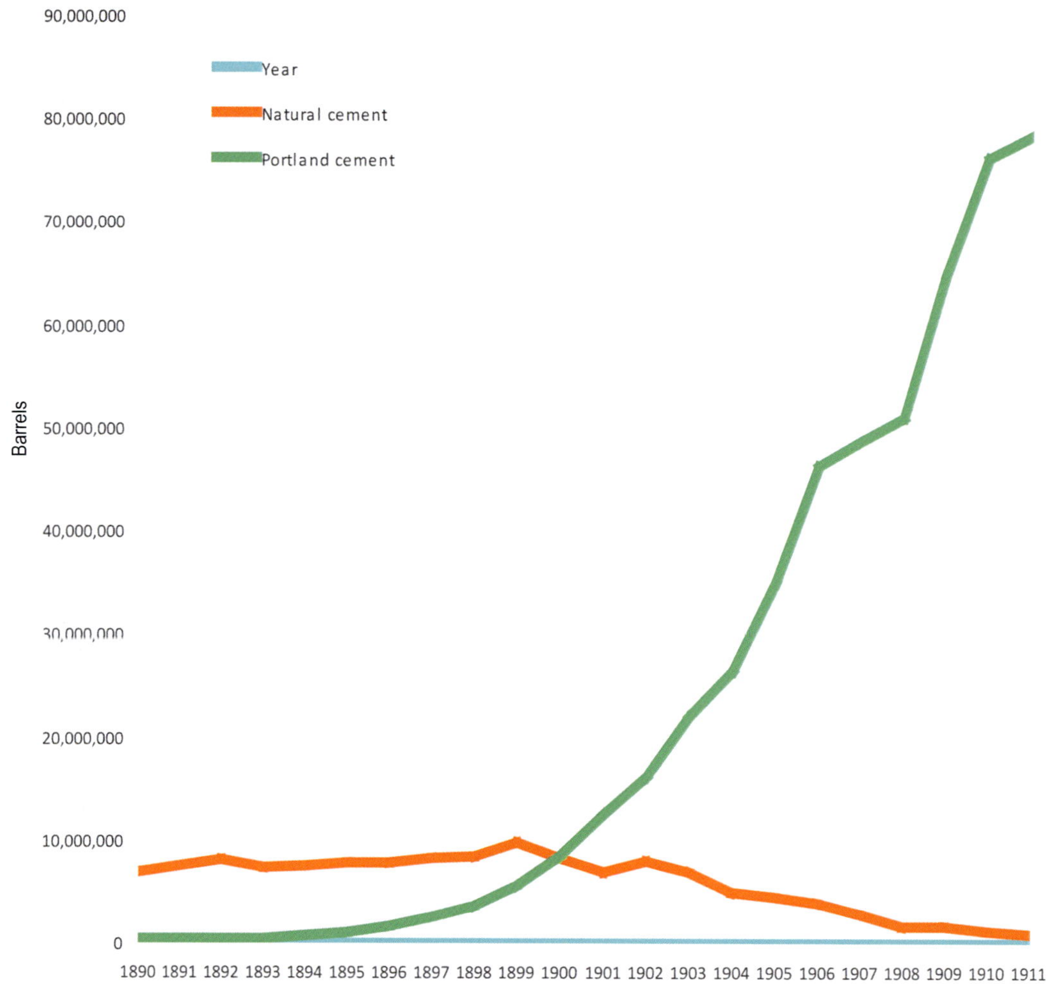

Fig. 335. Growth of the cement industry in the United States between 1890 and 1911. During this time interval, natural cement production leveled off and then declined while the use of Portland cement experienced astronomical growth in part by replacing the use of natural cement, but primarily through the rapidly expanding use of cement in concrete. The largest of these new uses included construction of road pavements, as motor vehicle traffic increased, along with the use of concrete blocks and poured concrete in general construction, thus changing thousands of years of building stone use. After Burchard, 1912.

it became more economical to manufacture Portland Cement and ship it to more distant markets. In a few short years of the late 1890s, the use of Portland cement skyrocketed while natural cement rapidly declined (Fig. 335). More specific to this story, Milwaukee Hydraulic Cement was driven off of the market.

Geology and mineral resources

The availability of mineral resources is controlled to a large extent by geological factors, which is demonstrated well by the development of the natural cement industry of the Milwaukee region. These factors including the depositional character of the bedrock, the persistency of these characteristics regionally, the structural configuration, and elevation of individual beds, along with the character and thickness of the overburden, all impacted the availability of cement rock in this area during the 19th century. Most important for any mineral resource is the character of the rock itself. Having the specific chemical composition in the correct proportions to make good natural hydraulic cement was the primary control on the value of this resource. Not all of the Milwaukee Formation made good natural hydraulic cement, however, and although the formation is over 60 feet thick, only its lower 21 feet could be used as cement rock. Most of the upper beds of the Milwaukee Formation are still only known from deep subsurface borings or temporary rock excavations which all show that there is significant stratigraphic variation within the formation (Raasch, 1935; Mikulic & Kluessendorf, 1988a; 1988b, 1998; Kluessendorf, Mikulic & Carmen, 1988). The cement rock strata of the Milwaukee Formation (Beds 1–11 of Teller, 1901; A and B zones of Cleland, 1911), which are equivalent to the Berthelet Member of Raasch (1935), are limited to the bottom 21 feet of the formation and were only exposed in their entirety when the cement quarries were in operation. Even with the correct composition, however, it was fortuitous that a few feet of these strata which were historically exposed in limited natural outcrops were recognized as a cement rock source and were later commercially quarried for that purpose.

Most of the property along the Milwaukee River from modern day Capitol Drive northwest to Hampton Avenue was eventually purchased by the Milwaukee Cement Company, but for geologic and economic reasons, only a limited area of this land was ever quarried or mined. Quarrying did occur where there was a thick deposit of cement rock at or near the bedrock surface which was exposed at or near the river bed in the Milwaukee River valley. This would combine a maximum thickness of cement rock with a minimal thickness of overburden making the quarrying of cement rock relatively easy and economical. In general, the cement rock of the Milwaukee Formation underlies a large area in northeastern Milwaukee County, but usually is deeply buried by Quaternary age unconsolidated glacial and lake sediments. The area around Estabrook Park presents a unique situation however, as the Milwaukee River has locally cut through tens of feet of unconsolidated sediments down to the bedrock surface exposing the top of a thick interval of cement beds which could then be quarried. The bedrock surface here is actually higher than is typical for much of the region and marks the location of a buried bedrock hill that is rising towards the northwest. Further complicating this geology but representing a key factor in the availability of the cement rock is a structural feature that has tilted the normally horizontal rock strata downward to the southeast in a syncline and upward to the northwest (Mikulic & Kluessendorf, 1988a). This relationship can be observed by plotting the changing elevations of the cement rock through the park area. At the southern end of the park near Capitol Drive, the rock surface was never exposed in outcrop and is deeply buried (100ft.). Although the cement rock found here has a thickness of over 20 feet, economic mining at that depth was beyond the technology of the time. At the "falls" further upriver, the top of the 20-foot-thick cement rock has risen about 80 feet and is at river level but covered by about 10 feet of overburden away from the river. Continuing up westward there is an interval where no rock is exposed along the river but at the Estabrook Park dam site the Thiensville Formation, which underlies the Milwaukee Formation cement rock, outcrops at the bedrock surface which is about 5 feet higher than at the "falls". Nearly all of the cement beds have been eroded away here but they formerly occurred at an elevation 20 or more feet higher than the current bedrock surface. Away from the river 10 or more feet of over burden overlies the rock surface. From the dam area northwest past the Port Washington Street bridge, these rock units change very little in elevation and with little or no Milwaukee Formation remaining, the Thiensville Formation generally occurs at the bedrock surface. Although the cement beds show considerable changes in elevation over this distance, their consistent uneroded thickness and their generally consistent lithologic character makes it an excellent source rock for cement manufacturing.

Geologically the Milwaukee Formation cement beds were deposited on a relatively level shallow sea floor during the Middle Devonian Period (390,000,000 years ago), which accounts for their consistence in composition. The 100-foot change in elevation of these beds between Capital Drive and the Estabrook dam site was the result of more geologically recent tilting of the bedrock and these beds were uncovered later through post Ice Age erosion by the Milwaukee River. From a mineral resource perspective, these beds underlie many tens of square miles in this part of Milwaukee County, but this is the only location in the region (from the southern end of Mill # 2 north towards the "falls" in Estabrook Park) where a significant volume of cement rock could be economically quarried during the late 19th century for the manufacture of natural cement.

The origins of the Milwaukee cement Industry

The discovery of Milwaukee's natural cement rock began with geological research starting in the 1830s. Increase Allen Lapham, Wisconsin's first resident scientist (Hays, 1995; Bergland & Hayes, 2014), had begun to study the local geology of the area when he moved to Milwaukee in 1836 (Mikulic, 1983). Previously he was employed as a construction engineer on the Louisville and Shippingport Canal and had at the age of 17 published a paper (Lapham, 1828) on the geology of the canal area around Louisville, Kentucky. In this paper, Lapham described the characteristics of several rock units exposed in the vicinity of the canal, one of which he identified as "water lime rock" which when calcined forms a cement "which has the property of setting very hard under water" (Lapham, 1828). In an unpublished original manuscript titled *Geology of Milwaukee County* written between 1844–1848 (Lapham papers, Wisconsin State Historical Society), Lapham described both Lake Michigan shore exposures and Milwaukee River outcrops as belonging to the same stratigraphic unit and commented on their natural cement potential (Mikulic & Kluessendorf, 1998). In describing these Milwaukee rocks, he noted "The color, fracture, texture of this rock, as well as many of the petrifactions are the same as that of the Water limestone at Louisville, Ky; and its geological position is the same, being next above the ["Cliff"] geodiferous limestone." On a following page he further stated, "Some layers of this limestone upon being burnt prove to be hydraulic lime, would not slack but when pulverized and made into a paste would harden under water as readily as the best hydraulic cement from New York." [This last page is undated, seems to be in different handwriting, and may have been a later addition to the manuscript]. Lapham published his first stratigraphic section of the Wisconsin Paleozoic rocks in 1851, where he correlated the limestone found along the Milwaukee River located north of the community named Humboldt and the Lake Michigan shore outcrops at Whitefish Bay with those of the "corniferous limestone of Eaton" in New York. In discussing the corniferous limestone, he stated, "Hydraulic limestone is found associated with it at Louisville and other places" (Lapham, 1851). In his last geologic publication (Lapham, 1876), he stated that "some of the layers appear to be of the same age as the water-limestones of Louisville Ky., and may have the same useful quality."

When in 1873 Lapham became the Chief Geologist of the Geological Survey of Wisconsin, a special effort was undertaken to investigate the cement potential of these rocks (Chamberlin, 1877, 1878). It is clear that Lapham had long known that these Milwaukee area rocks had the same basic characteristics of other Devonian age hydraulic natural cements already in production. It wasn't until the 1870s however, that Joseph R. Berthelet, Sr. also recognized the natural cement possibilities of the rocks along the Milwaukee River and decided to go into the cement business. Berthelet was co-owner of H. Berthelet & Company, a Milwaukee manufacturer of cement sewer pipe, and had been using hydraulic cement manufactured near Louisville in his business (Greene & Berthelet, 1949). Familiar with the characteristics of the rocks used to make Louisville cement, he was able to recognize that a locally collected rock sample he saw in the Milwaukee city engineer's office had the same appearance. He further realized that if this specimen represented a local source of cement rock he could save much of the transportation costs incurred by shipping Louisville cement to Milwaukee for use in his business. The often-cited story that the rock specimen Berthelet saw was from excavations for the North Avenue bridge in Milwaukee (Berthelet, 1942; Greene & Berthelet 1949) is untrue. It is now known that the bedrock surface is too deep (80 feet) at the bridge site to have excavated any rock, and that the specimen Berthelet saw was a rejected sample from the Milwaukee River outcrops to the north which had been submitted to the city by a contractor for possible use in the bridge's masonry. Exploring along the river Berthelet found and sampled the outcrops of the rock that Lapham had earlier studied and began a series of tests to determine if the rock would make a high-quality natural cement. Satisfied with his results, he then had Don J. Whittemore, chief engineer of the Chicago, Milwaukee, and St. Paul Railroad, conduct additional tests that eventually numbered nearly three thousand over the following year and a half (Langson, 1877). Whittemore was able to further confirm that the Milwaukee cement rock made high quality cement and determined, "that the cement produced from this stone is far superior in all the essential qualities of hydraulic cement to any other article of the kind in use at the present time" (Langson; 1876). Whittemore had previously (Whittemore, 1875) presented an extensive report on his work and stated:

"In conclusion, it gives me great pleasure to say to you, that this remarkable deposit of natural wealth is apparently abundant; that it lies in the bed and along the banks of our Milwaukee river, in the immediate vicinity of Washington bridge, not over a mile and a half from our city limits, in close to one or more of our trunk railway lines, and occupies a region of the river particularly noted for the beauty of its natural scenery. I congratulate you, moreover, upon the probability that, under the auspices of the enterprising gentlemen residing in this city who own the property, that this rock is likely to contribute largely to the general prosperity of Milwaukee." He further proposed to name this cement "Milwaukee Hydraulic Cement."

The Milwaukee Cement Company

Even before testing was completed, Berthelet and his associates including John Johnson, Henry Berthelet, and George H. Paul, began to purchase land along the Milwaukee River in order to open a quarry and erect a cement plant. What had started as an opportunity to reduce the expense of importing the Louisville cement Berthelet used in his cement pipe business, had rapidly become a plan to supply cement for the entire Milwaukee region. To finance this project, he gathered several key investors who eventually raised a capital of $350,000. Throughout much of its existence, many of these same individuals would remain connected to the company as investors, directors, or employees including: Joseph R. Berthelet Sr., Joseph R. Berthelet Jr., George H. Paul, John Johnston, Henry Berthelet, Don J. Whittemore, William T. Berthelet, Thomas A. Greene, Howard Greene, Henry Campbell, George S. Bartlett, William Plankinton, and Samuel Marshall. Of these individuals, Thomas A. Greene was one of the more interesting although he initially had no connection to the company. As co-owner of Greene and Button, a prominent Milwaukee area wholesale drug business, Greene was a successful local businessman (Greene & Berthelet, 1949). He also was a prominent local gentlemen naturalist, with a long interest in geology. Late in life Greene began to collect local fossils as a doctor-ordered way to relax from pressures of business (Mikulic, 1983; Kluessendorf & Mikulic 1993; Kluessendorf & Mikulic, 1997; Mikulic, 1991). His attention quickly focused on the newly opened Milwaukee Cement Company's quarries, which were the only significant source of Wisconsin Devonian fossils. Becoming a regular fossil collecting and buying visitor to the quarries, he eventually purchased stock in the company (probably as a way of securing access to the quarries), and later became one of its directors. His collection of the cement quarry fossils is probably the largest and one of the most important ever made and is now part of the Greene collection at the University of Wisconsin-Milwaukee. After his death in 1894, his son Howard Greene took over his father's position in the cement company continuing a multi-generational family association with that institution like that of the Berthelet family. Long after it had ceased producing cement, Howard Greene and William Berthelet served as the last officers of the company. Through most of the existence of the Milwaukee Cement Company, members of the Berthelet family also continued to operate H. Berthelet & Co. in an adjacent office to that of the Milwaukee Cement Company office on Water Street in Milwaukee.

Fig. 336. Letterhead of the Milwaukee Cement Company circa 1890. View on left side shows Mill # 2 (left) and Mill # 1 (right). Note illustration of the prominent Milwaukee Formation fossil cephalopod *Gyronaedyceras eryx* between "Milwaukee" and "Cement" on right side, which was probably done at the suggestion of T.A. Greene.

On November 25, 1875, a charter was granted by the State of Wisconsin to the Milwaukee Cement Company and construction of the plant and kilns began in early 1876. Later that year the company began a three-decade period of almost continuous growth in which its production regularly increased and its operations were frequently modernized. At first sales of its cement focused on the Milwaukee area market, but as the company became a success and the superior qualities of its product were widely known, sales increased throughout the Midwest. For about 25 years the Milwaukee Cement Company remained highly profitable, regularly paying large dividends to its investors. This resulted in the company being regularly featured in local business promotional publications, and described in national building materials and engineering journals. At various times, it was noted to be the largest cement plant in the country and among the most modern in existence (Lewis, 1898). During the late 1800s and early 1900s, there also were numerous book, magazine, and newspaper articles describing success of the company and its operations (see Langson, 1876, 1877, 1879; Land, 1883; Barton, 1886; Langson, 1891; Bartlett, 1892; Lewis, 1898 [reprinted 1900]; Brown, 1901; Peckham, 1901). The 1949 historical article by Greene & Berthelet, reprinted with numerous illustrations and background information in *Natural News*, 2006a, 2006b, 2007, covers many of the business and administrative operations of the company that will not be repeated in detail here.

Fig. 337. The original plant (later Mill # 1) of the Milwaukee Cement Company showing the quarry in the foreground, two kilns in right center of photograph beyond the quarry, and the mill to west (left) of kilns. This view is looking north from the original south bank of the Milwaukee River. The plant was located in what is now Estabrook Park and east of the former railroad bridge site. The stage of kiln development indicates that this photograph was taken in the summer or fall of 1876. The photograph is part of a series by J.C. Miller including Figs. 339 and 340 in this publication. Used with permission of the Wisconsin Historical Society, WHS-142690.

The actual construction of its operations was started around May 24th, 1876, when workers "were directed to begin clearing the bedrock surface of overburden on a 50 by 100-foot area between the kiln site and river" in preparation for the quarrying of cement rock (Lohr, 1964). Construction of the adjacent kiln and mill (later Mill # 1) had begun earlier in the spring of 1876 (Greene & Berthelet, 1949). This plant and quarry area were located on the east side of the Milwaukee River, south of both the "falls" and rail spur (the course of the river generally runs from northwest to southeast through this area, but is usually described as running north–south) (Fig. 337). The "falls" is not a natural feature but the location of the north wall of the original quarry. The rail spur was built to the mill and kilns from the nearby Chicago & Northwestern Railroad (Northwestern Union) line providing a convenient method of shipping cement from the plant. A contemporary newspaper article (Milwaukee Daily Sentinel, June 23, 1876) provides a fairly detailed description of the initial operations at the company's quarry and plant (later Mill # 1) along with some important geological observations. "About 100 feet west of the mill a quarry had been opened and there a dozen men were busy in digging out the stone. About four feet of surface clay was removed and the unbroken bed of rock lay bare". The rock was in "layers about 2 feet thick, and blasting powder is necessary to loosen it." After blasting "the stone is broken into pieces that a man can handle with ease" and loaded into quarry carts (Figs. 338 and 339).

The carts were then hauled on a tramway leading up to the top of the kilns, only one of which was then in operation with another under construction. The kiln was square, 40 feet high, 22 feet long and 20 feet wide with an egg-shaped interior that was 7½ feet in diameter at the top, 10 feet in the middle and 3 feet at the opening at the bottom (Milwaukee Daily Sentinel, 6-23-1876). Although the type of the materials used in kiln construction was not described, contemporary photographs (Figs. 337 and 340) indicate the outer walls were probably built with large dressed blocks of cement rock, while the internal lining of the kiln was probably made of fire brick. During operation, the kiln was filled regularly at its top with alternating layers of cement rock and coal. After continuously burning for twelve hours, the burned stone (clinker) had sunk to the bottom of the kiln and was ready for removal and grinding. From the kiln the clinker was transported on another tramway up into the mill building where it was processed through crushers before dropping down to the mill stones where it was ground into a fine powder. From there the powder was passed to a packing room where it was either placed in barrels or bags which were stored in a large warehouse before shipping to customers. This would be the basic system used by the company over the next twenty-five years, but it was regularly expanded and modified as its output rapidly expanded to meet increasing demand and new production technology was developed.

The Milwaukee Cement Company became well known as an innovator in the manufacture of natural cement and its plants were long considered to be the most progressive in the country. At the time of its opening, the mill turned out about 100 barrels of cement a day operating with a single kiln but by the end of the year it had produced 20,000 to 30,000 barrels of cement with 3 kilns in operation (Lanyon, 1877). After the 1876 opening of the original plant, expansion and innovation became a regular feature of the company. For example, in 1878, The Milwaukee Cement Railway Company was formed to extend the Chicago & Northwestern Road rail spur from the Milwaukee Cement Company plant west across the Milwaukee River to connect with the main line of the Chicago, Milwaukee, & St. Paul Railway. The noteworthy feature of this 1.19-mile extension was the construction of a bridge across the Milwaukee River having "three spans of ninety feet each, which with the trestle-work approaches, will give the structure a length of about 400 feet" (Milwaukee Daily Sentinel, 8-7-1878), which provided the company excellent connections with the regional transportation network. This short route was still in use by other firms into the 1990s when the bridge was finally removed and the line abandoned. Other early developments included a fifth kiln being erected which, in contrast with the earlier kilns, was circular in shape and encased in boiler iron; a style that all future company kilns would have, and a third warehouse was built with production reaching 12,000 to 15,000 barrels per month (Langson, 1879). To improve communication, a five-mile-long private telephone line was built in 1879 to connect the company's Water Street office in Milwaukee with its plant on the Milwaukee River (Berthelet, 1942).

Fig. 338. An illustration from Chamberlin (Plate XIIIA, 1877, 1878), showing quarrying at the Milwaukee Cement Quarry in 1876. Included are depictions of the stripping of 5 to 10 feet of overburden above the cement rock, drilling and loading holes in the quarry wall in preparation for blasting, breaking blasted rock, and loading the rock into dump cars which were probably pulled by horses on a tramway to the kilns. The illustration also shows the significant increase of overburden that needed to be stripped as quarrying progressed away from the river and into the bluff. The mechanical structure on the left side of the illustration is probably a man operated water pump needed to keep the water level low in the quarry. This illustration was made by the Milwaukee Lithography & Engraving Company, and is based on a photograph (Fig. 339) taken by J.C. Miller during the summer or fall of 1876.

Fig. 339. The original 1876 photograph of the Milwaukee Cement Company quarry used as the basis for the illustration in Fig. 338. Photograph by J.C. Miller and used with permission of the Wisconsin Historical Society, WHS-142688.

Fig. 340. Photograph of the original Milwaukee Cement Company plant (later Mill # 1), showing the kilns (on left) with the mill and engine house (on right). Photograph by J.C. Miller and used with permission of the Wisconsin Historical Society, WHS-142686.

From a geological standpoint, it is interesting that while the cement rock had been extensively tested for over a year, information in this article indicates that the total thickness of the cement rock had not been determined when production started. Although the company had dug a few exploration holes, some of which were "16 feet deep," none had reached the bottom of the cement rock. The 1876 view of the quarry (Fig. 338) published in Chamberlin (1877, 1878) and original photograph on which it was based (Fig. 339) indicate that only about 10 feet of cement rock was then being quarried. At the time, the large property holdings of the company along with the modest planned scale of production suggested to many that the cement deposit was "inexhaustible" (Milwaukee Daily Sentinel, 6-23-1876); however, the rapid expansion of production in combination with the unrecognized geological limitations of the deposit would result in future supply problems.

The rapid success of the Milwaukee Cement Company probably attracted the early interest of other parties hoping to open competitive local cement plants even before any cement had been produced (Milwaukee Daily Sentinel, 6-23, 1876). However, the geology of the region greatly limited the occurrence of accessible deposits of cement rock. To try to prevent competition and insure their own supply of cement rock, the Milwaukee Cement Company attempted to purchase or control all of the land along the Milwaukee River where cement rock outcropped. After their initial purchases, the largest area of accessible cement rock remaining outside of their control was owned by the J.J. Orton estate. This property covered almost ninety acres on the west side of the river, and much of the land had a thick deposit of cement rock with only a thin covering of overburden, making it the best location for another quarry. By the mid 1880s the rapid growth of the market for Milwaukee Hydraulic Cement resulted in the company looking for ways to increase production. The availability of cement rock to quarry was a key factor in increasing production and it could only be addressed by acquiring more land. The most obvious source of additional land was the Orton property. When the company attempted to start negotiations for its purchase, it was discovered that the Orton estate had already received an offer for the property from the newly formed Hadfield Cement Company. This company was incorporated by the Hadfield family of Waukesha on December 19th, 1885, even though they had not previously been in the business of manufacturing cement. The Hadfields were however one of the largest producers of lime and building stone in the Midwest and were actively trying to expand their business interests into other ventures such as coal and apparently manufacturing and selling cement. Although the Hadfield company was first to make an offer on the Orton property, it was not accepted and a much higher bid by the Milwaukee Cement Company was able to secure the site (Greene & Berthelet, 1949). In 1887, the company built its Mill # 2 (Figs. 341 and 342) and opened a new quarry on the Orton estate land. Initially this plant was operated on a royalty basis, but in 1890 the entire property was purchased outright by the company. With twenty kilns and a potential production of 2,500 barrels of cement daily, this plant had a larger output than Mill # 1 and was considered to be the largest individual cement mill in the country (Stickney, 1897). Lewis, (1898) described Mill # 2 in a detailed article and although it had a similar layout to the original 1876 plant, it was much larger and utilized a number of significant technological improvements which greatly increased its production (Fig. 343). Several of these improvements resulted from the work and designs of Milwaukee Cement Company personnel. Most notable were the Berthelet separator and the Campbell kiln (anonymous, 1897; Lewis, 1898) developed by J.R. Berthelet and H. Campbell, respectively. They were issued on patents on their inventions, which were also used in other cement plants around the country. With frequent improvements in production and as one of the newest cement plants in the country (Stickney, 1897), the Milwaukee Cement Company continued to be considered to be a leader in cement manufacturing.

The purchase of the Orton land provided an important opportunity for a major 1892 change in the Milwaukee Cement Company's operations, the results of which are still a prominent feature of the area today. After years of excavation, the Mill # 1 quarry only had limited space available for expansion, so as far back as 1880, the company had to undertake more expensive and less efficient underground mining to meet its production needs. By 1886 several tunnels had been dug into the east wall of the quarry lying between the kilns on the south and the railroad bridge on the north. No mine maps of this site are known, but the tunnels might have extended to the company's eastern property line. With the purchase of the Orton land, the quarrying of cement rock at Mill # 1 was shifted west across the river. This was accomplished by diverting the Milwaukee River eastward using the quarry of Mill # 1 as a new artificial river channel. The modern "falls" in the park at the north end of this channel is not a natural feature, but is the north end of the Mill # 1 quarry representing a manmade artificial waterfall. In 1893 Mill # 1 was remodeled which included building a combination iron trestle bridge and inclined tramway to the kilns across the newly rechanneled river (Figs. 343 and 344). The tramway provided a means to move all of the cement rock used in the kilns from a new quarry along the now dry old west bank area of the river. With the prior opening of the Mill # 2 quarry, all the cement rock quarrying was shifted to the original west side of the river (Figs. 343 & 344) and eventually both of these new quarries were joined making one large quarry which extended from Mill # 2 north past the railroad bridge.

In 1898, an article on changing the river course appeared in Stone magazine which was based on a story in *The Evening Wisconsin* newspaper.

"One of the most interesting things that has been done at the cement works is the changing of the course of the Milwaukee River for a distance of over half a mile. To the east of the old course of the river the rock was taken out and the quarries worked as far as they would pay. Then it was concluded that immediately west of the river was a fine bed of rock that could be worked to advantage if the river were out of the way. So, a dam was built about three-quarters of a mile below the Washington bridge and the river narrowed down to about half its former size. Then the water was sent through the worked-out quarries to the east of its original course. An embankment was raised between the old and new courses and now the old river

Fig. 341. Lithograph depicting Mill # 2 (on left) and Mill # 1 (on right) of the Milwaukee Cement Company. View is from the east side of the Milwaukee River looking towards the northwest. Note that the Mill # 2 quarry is limited in size and the river has not been directed into old quarry at Mill # 1 indicating the time period of this illustration was between 1887 when Mill # 2 was started and 1890 when the river was diverted. Courtesy of Historic Photo Collection / Milwaukee Public Library.

Fig. 342. Panoramic photograph showing, from left to right, the Milwaukee Cement Company Mill # 2, the quarry for Mill # 2, the Milwaukee Cement Company Mill #1, and the plant of the Cream City Cement Company. This view is from east side of Milwaukee River looking southwest. Note that the extensive quarry expansion at Mill # 2 (started 1887) and the presence of the Cream City Cement plant (demolished after 1894) suggests that the date the photograph was taken was in the early 1890s. Courtesy of Historic Photo Collection / Milwaukee Public Library.

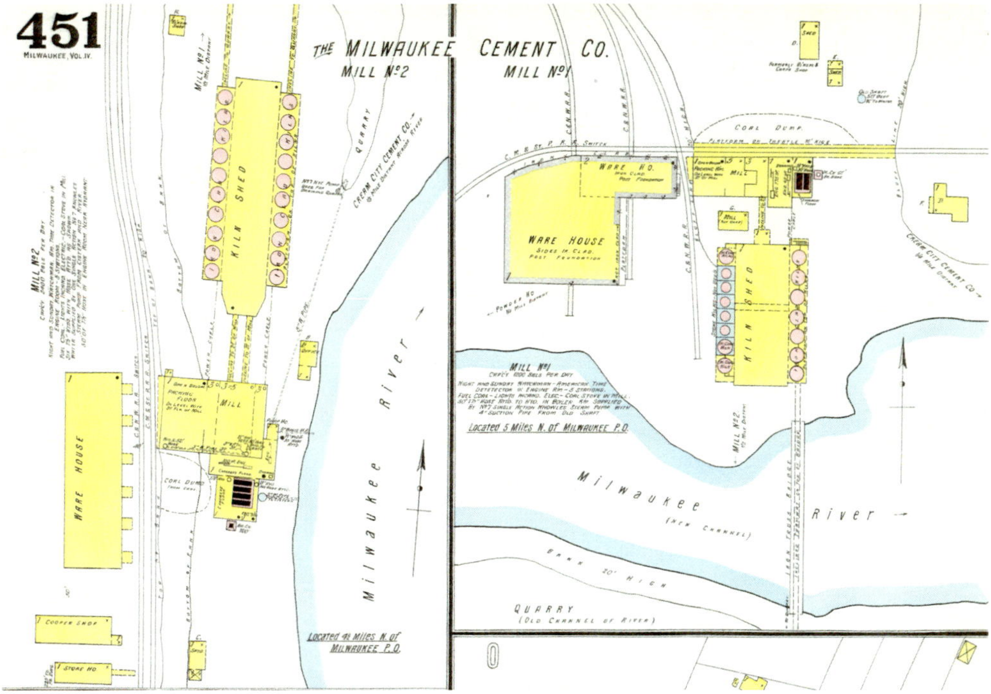

Fig. 343. Fire insurance maps of the Milwaukee Cement Company's Mill # 1 and Mill # 2 showing the layout of each plant and the primary materials used in the construction of buildings and other structures. From Insurance Maps of Milwaukee, Wisconsin, Volume Four, 1894, Sanborn-Perris Map Co., p. 451. Pink color indicates buildings primarily of brick construction, yellow is wood construction, and blue is stone construction. From the American Geographical Society Library, University of Wisconsin-Milwaukee Libraries.

Fig. 344. Milwaukee Cement Company Mill # 1 showing the remodeled 1893 plant, along with the new bridge and trestle across the Milwaukee River which had been rerouted through former Mill # 1 quarry. This view is looking west and possibly was taken from a building at the Cream City Cement Company plant. From a photograph in Views of the Cement Co.'s Works, Milwaukee, Wisconsin by the Milwaukee Cement Company, 1897. Used with permission of the Wisconsin Historical Society number, WHS 142689.

bed is perfectly dry. This change was made about a half a dozen years ago. Quarrying was commenced at the southern end of the dried-up bed and the stone has been taken out for a distance of about an eighth of a mile, and to a considerable depth." "In the neighborhood of the cement works the river is shallow for the most part, except where it runs through places where rock has been removed."

With both quarries and plants running, the Milwaukee Cement Company set new records of production yearly. Following the construction of Mill # 2 in 1887 and the remodeling of Mill # 1 in 1893, the company had thirty kilns in operation producing up to 500,000 barrels of cement and employing about 300 workers. Technological innovation and regular modernization (Figs. 345 & 346) resulted in one of the most productive cement operations in the country. By 1893 the Milwaukee Cement Company Mill # 1 had 10 kilns in operation and the company's operations were claimed to be the largest, most modern and complete hydraulic cement works in the world (Fig. 347). The company continued to have exceptional growth into the 1890s, with over 100,000 barrels produced by Mill #1 in 1880, over 200,000 barrels in Mill # 1 in 1884, over 300,000, barrels from Mill # 1 in 1887, and over 400,000 barrels of cement from 30 kilns in Mills # 1 and 2 combined in 1888. In 1896 more than 500,000 barrels of cement were produced in Mills 1 and 2, which might represent the company's largest production in a single year. When in 1888 Mill # 2 began operation, it was stated that it had a daily capacity of 2,000 barrels a day, which could make the yearly production for both plants at over a million barrels which "is far in excess of the demands" (Milwaukee Sentinel, July 7, 1888). Although production increased until at least 1896, the rate of increase slowed in the 1890s, possibly because of a weak economy following the Panic of 1893, but certainly because of the start of competition from a rapidly growing Portland cement industry.

Competition

After the Orton property was purchased, nearly all of the best locations for quarryable cement rock were owned by the Milwaukee Cement Company. There were other spots where cement rock outcropped in Milwaukee County which were not ideal geologically, but because of the high demand for and potential profitability of manufacturing Milwaukee Hydraulic Cement, they attracted the attention of other groups. Over time two other companies built cement plants, but because of the geological conditions at their locations, neither was successful.

The first was the Cream City Cement Company, which was incorporated in 1889 by a group of Milwaukee investors including George Brumder, a prominent local German newspaper publisher. The company built a large plant with five kilns on a piece of land along the east bank of the river which was so small that it apparently had escaped the interests of the Milwaukee Cement Company. Howard Greene (unpublished and undated manuscript) stated that the reason why the company hadn't purchased it earlier was that Berthelet didn't think the land had any value. This mistake would cause the company a lot of trouble. Located near the middle of the Milwaukee Cement Company property, this new plant site consisted of only about a three-acre triangular piece of land mostly located on the top of the river bluff between the Chicago and Northwestern Railroad tracks and the river. While this location had a good section of cement rock to work with, other significant geologic constraints made it difficult and expensive to do so. The site included virtually none of the river flood plain area that provided easy access to the cement beds while the land here was covered by more than 40 feet overburden. This parcel was part of a larger property of about 20 acres in total that extended to the north, none of which could be quarried from ground level, but it could be accessed by a vertical shaft with lateral tunneling extending from the plant site. To access its cement rock the Cream City Cement Company had to construct its kilns and plant on top of the bluff. It was then connected by a platform out and over the river flood plain to an elevator located over a mine shaft (Figs. 348 & 349). All of the cement rock produced on this site was mined underground and had to be transported up to the top of the bluff, which added significantly to production costs. The only description of what the inside of any of the local cement mines were like was given by Johanna Thomsen in a Milwaukee Journal article (Milwaukee Journal, 2-23-1960). Her husband was a worker at the Cream City Cement plant and in this article she remembered going down the elevator shaft to a tunnel where the men were working. The

Fig. 345. Steam engine used to pull dump cars filled with cement rock from the quarry working face to the kilns in Mill # 2. Originally horses had been used to pull dump cars on the tramway in the Mill # 1 quarry; however, when this photograph was taken, sometime after 1887, steam power had replaced horses. This is a small example of the modernization regularly undertaken by the Milwaukee Cement Company at its operations. Courtesy of Historic Photo Collection / Milwaukee Public Library.

Fig. 346. Stone dump cars and steam engine used to transport cement rock to the kilns in the Mill # 2 quarry sometime after 1887. Courtesy of Historic Photo Collection / Milwaukee Public Library.

Fig. 347. Portion of map of Milwaukee Cement Company's property showing area around Mill # 1 + Mill # 2. This map was probably made after 1894 by R.R. Reinertsen as the surveyor. The 3 acre parcel located east of Mill # 1 and north of Milwaukee River near center of map is the former location of Cream City Cement Company plant. Used with permission of Wisconsin Historical Society and University of Wisconsin-Milwaukee Archives, 00001-scan_2019-01-25_08-37-30; detail of full map.

light in the mine was supplied by kerosene torches and the floor was always wet and pumps were working night and day. Pillars were left to hold up the ceiling which was also shored up by timbers. The Cream City Cement Company operated for only a few years but was "a serious competitor of the Milwaukee Company by their policy of cutting prices and selling below their cost. This ultimately proved to be a ruinous policy" (Berthelet, 1942). In 1894, the company and its land was purchased by the Milwaukee Cement Company more likely to eliminate a competitor than to gain its reserves of cement rock and its "plant was wrecked and tunnels abandoned" (Greene & Berthelet, 1949).

The only other significant cement rock exposures in the Milwaukee area were at Whitefish Bay and located at the base of a steep bluff along the Lake Michigan shore. Although probably known since the 1670s, the site could not be easily accessed for any type of stone quarrying as the rock bottom of the lake at this point made the water too shallow (18 to 24 inches) to approach by boat for commercial shipping. With no way to ship cement on the lake and no room for kilns or a plant along the shore, quarried rock had to be transported up to the top of the steep 100 foot high bluff where the plant and kilns were eventually built adding to production costs. Lapham (1844, unpublished manuscript) commented, "The clay bank above the rock is so steep that the fisherman stationed here are obliged to use a ladder for a part of the ascent". However, the potential profitability for investors to open a new cement operation at Whitefish Bay was too great to ignore. On September 1, 1892, the Milwaukee Brick and Cement Company was incorporated by Milwaukee investors to manufacture hydraulic cement, mud and pressed brick, sewer pipes, and tiles on their 143-acre site along the Lake Michigan shore (Anonymous, 1892). In January, 1893, a description of the Whitefish Bay cement plant was published (Baxter, 1893), with an illustration (Fig. 350) of the planned operations. The plant was to be built at the top of the bluff and would be connected by an incline to a quarry in the lake bed. Cement rock would be quarried from the lake bed where a water tight dyke was to be built surrounding a large area of the bedrock surface extending 1000 feet out from the shore. This basin would be pumped out and then stone could be quarried from the former lake bed. It was proposed that shipping of both brick and cement would be by rail at the top of the bluff and possibly by boat following the construction of piers and docks on the lake. A brick plant was built in 1893 on the top of the bluff and for a number of years brick was produced from clay excavated there. However, because of the complexity and cost of excavating cement rock at the lake level and hauling it up to kilns and a mill at the top of the bluff, the start of this part of the business was postponed for many years. Another possible factor in this delay was the financial Panic of 1893 which resulted in a serious economic downturn that lasted for several years. In 1899 construction was finally started on the cement plant, and in 1900 the company name was changed to the Consolidated Cement Company of Milwaukee with a renewed effort to enter the cement business. As previously planned, the kilns and a mill were built at the top of the bluff, however the plan to dyke off a quarry in the lake bed was abandoned and underground mining at the lake shore was started instead, presumably as a less complicated and expensive approach. Alden (1906, 1918) described the operations here as:

"Near this point, the Consolidated Cement Company has opened a shaft to supply its mill upon the bluff above. This shaft extends to a depth of 22 feet below the level of the beach, where it connects with a main gallery which, at the time of the writer's visit, extended 210 feet in a north–south direction. From this gallery drifts were being extended to the west in the working beds. The upper 5 or 6 feet is left as roofing and is not well exposed to examination, but material thrown out in excavating the shaft through these showed them to be of moderately hard, fine-grained, bluish-gray hydraulic limestone, characterized by an abundance of pyritized fossils, crystals of pyrite and of calcite, and traces of bitumen. The working face, which has a maximum height of 14 feet, shows dense, bluish-gray, massive to heavy-bedded magnesian limestone, having the same general appearance as that at the quarries on the river, but less distinctly marked bedding planes. So far as could be seen this part seems to be less fossiliferous than the beds composing the roof and those exposed along the river. The bedding is slightly undulatory, so that, though the general dip of the formation is toward the southeast, there is a slight dip to the west in the northern part of the exposure."

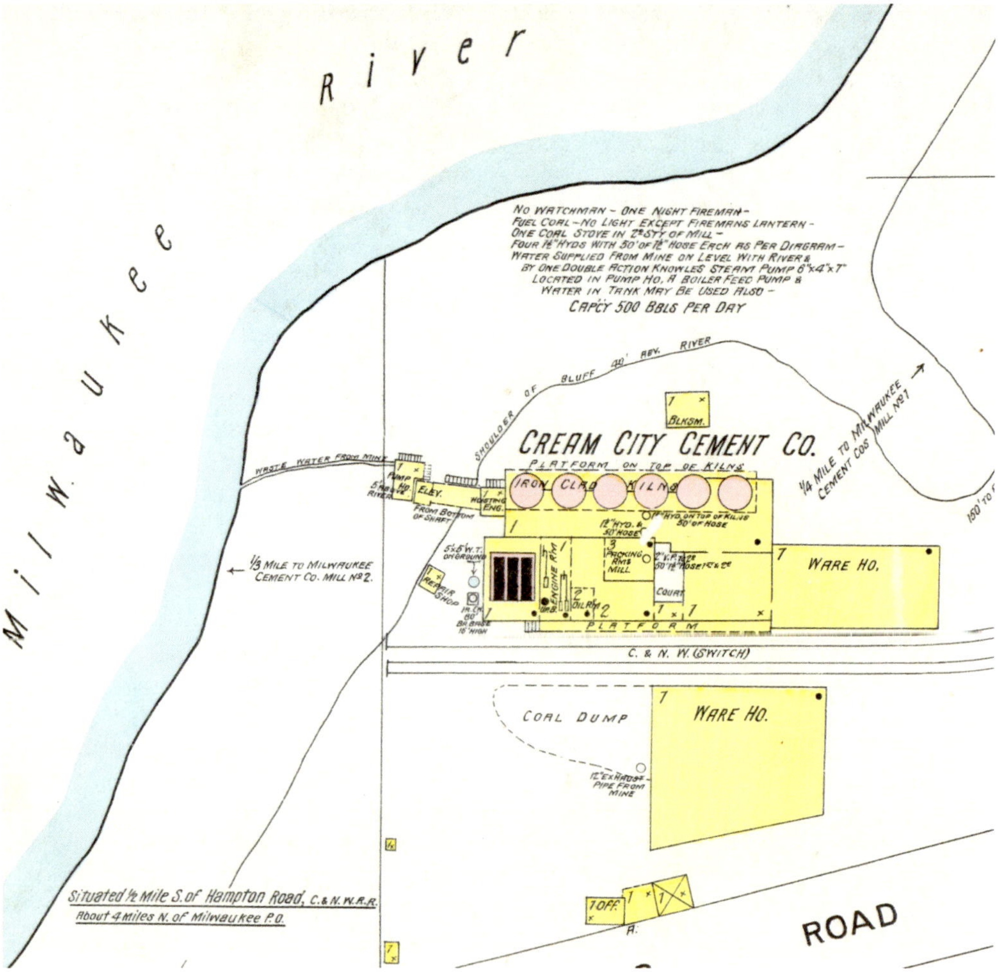

Fig. 348. Fire insurance map of the Cream City Cement Company showing the layout of the plant and the primary materials used in the construction of buildings and other structures. From; Insurance Maps of Milwaukee, Wisconsin, Volume Four, 1894, Sanborn-Perris Map Company, p. 459. Pink color indicates buildings primarily of brick construction, and yellow wood construction. From the American Geographical Society Library, University of Wisconsin-Milwaukee Libraries.

The roof rock was likely to be the same thick beds of Cleland's B zone and Raasch's Berthelet member which was also used for roof rock in the mines along the Milwaukee River. The cement operations of the Consolidated Cement Company were short lived. By 1904 the owners were asking for relief in their property taxes and in 1906 the company was no longer listed on local tax rolls (Nelson & Roberts, 1963), suggesting operations had ceased. In 1905, the operations were offered for sale in trade journals (Fig. 351).

Cleland (1911) stated that no work was being done at the mine when he visited the site in 1907.

Although a significant investment had been made to build a plant and kilns, construct a haulage system to bring cement rock to the top of the bluff, and develop a mine on the lake shore, the entire project was probably at the wrong place and certainly at the wrong time. The geologic conditions of the site limited the source of cement rock to underground mining at the lake shore, while also limiting the site of a plant and kilns to the bluff top where a transportation connection for rail shipping could be used. These complications by themselves would have made it difficult to manufacture cement as cheaply as was being done at the Milwaukee River locations of the Milwaukee Cement Company. However, the economic downturn resulting

Fig. 349. Right side of panoramic view in figure Fig. 342, showing the Cream City Cement Company plant with its mine shaft elevator on the right and the Milwaukee Cement Company Mill #1 quarry to the left. This view is from east side of Milwaukee River looking west. The development of the Milwaukee Cement Company Mill # 2 (started 1887) and the presence of the Cream City Cement plant (demolished after 1894) suggests that the date the photograph was taken was in the early 1890s. Courtesy of Historic Photo Collection / Milwaukee Public Library.

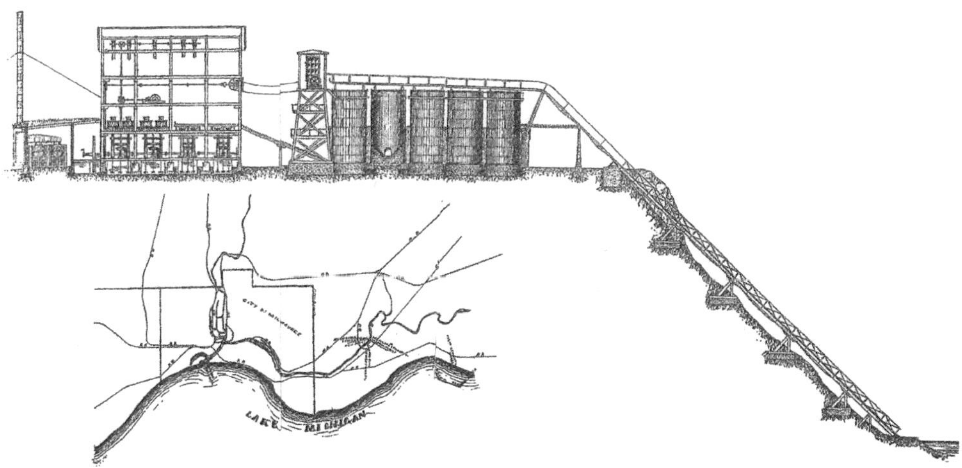

Fig. 350. Diagram showing the 1893 design of the Milwaukee Brick and Cement Plant to be built at Whitefish Bay. When the plant was finally built in 1899 under the name of the Consolidated Cement Company, the idea of quarrying cement rock in the lake bed had been changed to underground mining. This would have resulted in some changes to this diagram, but the basic concept probably remained the same. From A.E. Baxter, 1893.

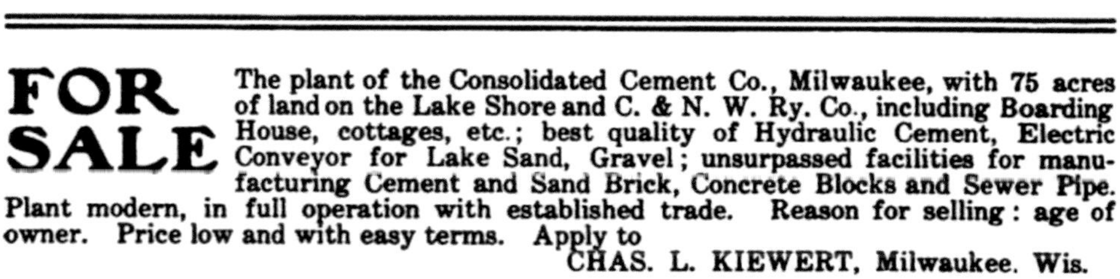

FOR SALE The plant of the Consolidated Cement Co., Milwaukee, with 75 acres of land on the Lake Shore and C. & N. W. Ry. Co., including Boarding House, cottages, etc.; best quality of Hydraulic Cement, Electric Conveyor for Lake Sand, Gravel; unsurpassed facilities for manufacturing Cement and Sand Brick, Concrete Blocks and Sewer Pipe. Plant modern, in full operation with established trade. Reason for selling: age of owner. Price low and with easy terms. Apply to
CHAS. L. KIEWERT, Milwaukee. Wis.

Fig. 351. Advertisement for sale of the Consolidated Cement Company in the July, 1905 issue of *The Cement Age*.

from the Panic of 1893 probably also had a role in delaying the development of the site, and reducing potential sales. This was followed by the rather abrupt collapse of the natural cement industry in general as Portland cement started to dominate the market ending any potential for further natural cement production the Milwaukee area. The Milwaukee Journal in a 1960s article stated that the mine shaft here was filled in 1929 although the partially water filled tunnels were left open. In recent decades when the surface level of Lake Michigan water was at its lowest, remnants of various cement manufacturing structures and equipment can still be seen along the lake shore (Lohr, 1964), but no other indications of the cement industry at this site are visible.

It is unlikely that either the Cream City Cement Company or the Consolidated Cement Company mines had any significant role in geologic research on the Milwaukee Formation or in supplying fossils for scientists and collectors. The natural lake shore exposures near the Consolidated Cement Company mine site have long been a somewhat less productive collecting site when compared with those at the large Milwaukee River quarries and have been partially covered by erosion protection structures in recent years.

The End of the Industry

The Milwaukee Cement Company had a rapid growth and an even more rapid decline. Despite having an excellent source rock for manufacturing natural cement, operating modern and innovative plants, and had an extensive sales network covering most of the central United States, the company, which was one of the dominant cement producers in the country, was powerless to stop its own collapse. Howard Greene and William Berthelet, who were prominent in the Milwaukee Cement Company, described (1949) the collapse as follows:

"Just when the company was at the height of its activity something occurred that sounded the death knell of the natural cement industry. The rotary kiln was invented. Through its use, what is known as Portland cement could be produced at low costs from various types of materials. Portland cement develops its early strength much sooner than natural cement, and also withstands climatic conditions much better."

This change from natural cement to Portland cement was even more dramatic than it sounds. In the 1890s and 1900s the demand for and production of cement skyrocketed (Fig. 335). Part of this expansion resulted from new uses of cement in general, as Portland cement could be used in applications that natural cement wasn't well suited for or had previously existed. Portland cement for example replaced much of the use of lime mortars as a cementing agent. But more importantly was the major change in the types of building materials used in construction that occurred at the time. Cement used in concrete had begun to supplant stone materials that had been used in construction for thousands of years. Poured concrete and concrete blocks replaced building stone as the primary materials used to erect large structures. In conjunction with internal steel support it became possible to build true skyscrapers. The superior qualities of Portland cement resulted in new and expanded uses of cement in general. The problem with natural cement wasn't that it couldn't be used for some of these purposes but it simply wasn't as good or as cheap in many cases. Important new uses for cement also began at a time when economic prosperity and population growth would require vast amounts of cement to develop the infrastructure of a growing society. The most important of these new uses was in the form of road pavements needed for the transportation revolution following the invention of motor vehicles.

But if Portland cement hadn't supplanted natural cement, the Milwaukee Cement Company would probably have continued as an important cement producer; however, through no fault of its own and with no way to prevent the change, this prominent Milwaukee business rapidly declined. This was all due to the character of its cement rock which for a while was the best anywhere but was then replaced by a "new" product that was better for making concrete road pavements and concrete blocks. To keep its business going during the early 1900s, the company tried selling Portland cement which allowed the manufacture of its natural cement to continue on a small scale (Greene & Berthelet, 1949), but all production was finally suspended in 1911. A few of the other U.S. natural cement plants were able to continue low levels of production for a few more years, but the Milwaukee Cement Company was not one of them.

The Quarry Sites After the Close of the Industry

When the quarrying and mining of cement rock ended in 1911, the Milwaukee Cement Company owned about 350 acres along the Milwaukee River at the northern edge of Milwaukee (Greene & Berthelet, 1949). Although over the years many other quarries in southeastern Wisconsin had been temporarily closed for economic reasons but later reopened, the extensive property holdings of the Milwaukee Cement Company were permanently abandoned as a source of any type of construction stone. What for several decades had been some of Wisconsin's most valuable mineral resource land would see its value rapidly decline and the additional production of cement rock or other types of stone products would never occur at the site again. The reasons for this abandonment were due to both geological and technological factors. Geologically the dolomitic rock of the Devonian Milwaukee Formation made excellent natural cement but it could not be used in making Portland cement its replacement because of its fixed chemical composition along with its high magnesium content. While other quarries in the Silurian dolomite rocks of southeastern Wisconsin could be used to make lime, crushed stone, or building stone, Milwaukee cement rock contained too much clay to make lime, was too soft to use as crushed stone, and was too weak and thin or irregular bedded to produce good building stone. Although there were large areas of the quarries where all of the cement rock had already been excavated, digging deeper into the underlying Thiensville Dolomite wasn't an option because of its poor quality as a construction material.

After the production of cement was discontinued, the company's mills were leased to the Michigan Quartz Silica Company which ground quartz from a gold mine in Michigan to make wood filler, paint filler, and polishing abrasives (Berthelet, 1942; Hore, 1915). This company first used Mill # 1 in its grinding operations but it was destroyed in a fire in March 1910 and they then switched to Mill # 2 which also burned to the ground in September 1916. After both mills were destroyed the company wrecked the kilns and salvaged their iron and bricks ending any further use of the plants. The stone ruins of the base of the kilns at plant # 1 could still be seen into the 1930s (Toepfer, 1977) but they were apparently buried. If these kiln foundations are still there they would be an important candidate for an industrial archeological study and preservation.

With the mills and kilns destroyed and no prospects of ever producing cement at the site again, the company began to sell its land around World War I. The City of Milwaukee was expanding north and the property had considerable value in providing new locations for industrial development. In addition, there was also a strong interest in preserving much of the land as parkland. The abandoned quarries had become mostly water filled and the manmade rugged topography of quarry cliffs and artificial lakes provided a unique and scenic landscape. Boating, fishing, hiking, picnicking, swimming and even fossil collecting (Mikulic & Kluessendorf, 2001), became popular activities along the river even though much of the land was never publicly owned. Some of the quarry area also attracted interest as sites for municipal garbage dumps. With these conflicting interests, the land was sold piecemeal until the 1940s and its ultimate use would long be controversial. The first important sale was made in 1916 when most of the land on the east side of the Milwaukee River was purchased by the new Milwaukee County Park system to create Estabrook Park. While the east side of the river became parkland however, the west side including the water filled Mill # 2 quarry, then known as Cement Lake, did not. In 1924, the City of Milwaukee bought much of the Mill # 2 quarry area to use as a dumping site for garbage. This started a controversy concerning the land use of this area which lasted decades and has still not been completely resolved. Specifically, the use of the west side property for dumping was in direct conflict with the plans of various groups that had interest in developing park land on both sides of the river including the Cement Lake area. In addition, using Cement Lake as a major dumpsite had a significantly negative impact on the public use of Estabrook Park as it was located directly across the river from the park spoiling its veiwscape and adjacent river use. Other problems resulted from the dumping such as smoldering dump fires (Henkel, 1939), and trash floating down river plagued adjacent neighborhoods.

By 1929 there was considerable opposition from local citizens to continued filling of Cement Lake with garbage. Charles Whitnall, the architect of the Milwaukee County Park System and its urban greenways (Platt, 2010), long a proponent of the creation of river parkways, had become one of the leaders of this opposition. The City of Milwaukee really wanted dump sites but also claimed "safety" as a primary reason to fill Cement Lake. The vertical old quarry walls, the twenty to thirty-foot water depth, along with the open entrances to the partially flooded mines all made the area dangerous and it became notorious for drownings (Henkel, 1939). Many of these drownings resulted from the presence of submerged drop-offs into deeper water which were typical for many abandoned quarries. More unique to the site however was the partially flooded mines. In the summer, it was possible to enter some mines by boat, while during the winter when the flooded quarries were frozen over, some of the mine ceilings were high enough for people to walk into them; however, the ice rapidly thinned back into the darkness where there was open water all year long (Mikulic & Kluessendorf, 2001). Because of the drownings, by the late 1920s there was public pressure to make the area safe. Plans were made to fence off mine entrances, although they ultimately seem to have been just buried under fill. The opponents to dumping pointed out that most of the mine related drownings were not occurring at Cement Lake but in other "Blue Hole" areas. Even geologists expressed their opposition to dumping because it would bury the unique exposures of fossiliferous Devonian rocks.

In 1930, following opposition to the complete filling of the lake, a plan was agreed upon to landscape what remained of the Cement Lake area into park land, filling it to a shallow depth for safety, and create an adjacent west side parkway. This would have preserved some of the rock exposures in the area which contributed to its unique scenic beauty (Henkel, 1939). While this plan seemed to resolve many of the preservation issues, dumping continued at Cement Lake. In addition, further changes to the landscape occurred during the 1930s when a large and mostly undisturbed section of the river was dramatically altered by rechanneling both in Lincoln Park and adjacent Estabrook Park (Ziemke, 1935). Much of this work was done as part of WPA projects, which included the construction of the Estabrook Park dam. The river rechanneling had some scientific interest as it involved extensive rock excavations of both the lowest beds of the Milwaukee Formation and the upper Thiensville formation, both of which were seldom exposed.

Despite this temporary reprieve, Charles Whitnall died in 1949 and by 1952 the city had gone back to using the "preserved" part of Cement Lake as a garbage dump and burn site. Despite continuing public protests, the entire area was eventually buried under trash, eliminating all indications of what was once a prominent industry along the west side of the river except for small areas near the railroad bridge. Despite promises from as far back as the 1920s, the Cement Lake site has not yet become a park site and since the 1970s has been a parking lot for the University of Wisconsin-Milwaukee. With decades of garbage dumping and park landscaping, nearly all of the formerly extensive rock exposures in outcrops and quarry walls have been obliterated. Several studies (Mikulic & Kluessendorf, 1982, 1997) have proposed preserving some of the very few remaining exposures of the Devonian rocks along the river but little has been accomplished to date. What had once been a grand window into the geologic history of Milwaukee and Wisconsin, a demonstration of the importance of mineral resources in building a major Midwestern urban area, while also providing a scenic and unique recreational site for the community, has almost completely disappeared, never to return. Estabrook Park is now a nicely landscaped park. Few would ever know that the conspicuous falls on the river are the manmade remnants of a once great industry.

Bibliography

Alden, W.C. 1906. Milwaukee. Special Folio 140. United States Geological Survey Geologic Atlas, 12 pp.

Alden, W.C. 1918. The Quaternary geology of southeastern Wisconsin with a chapter on the older rock formations. U.S. Geological Survey Professional Paper 106, 356 pp.

Anderson, W.J. & Bleyer, J. (Eds.) 1892. Milwaukee's great industries. Milwaukee, 352 pp.

Anonymous. 1892. Prospectus of the Milwaukee Brick and Cement Co. Incorporated. September 1st 1892, 8 pp.

Anonymous. 1893. Milwaukee, Wis.: The convention city for 1893: Milwaukee Brick and Cement Company. Street Railway Journal, 9: 639.

Anonymous. 1897. The Berthelet Separator and the Campbell Kiln in Cement Manufacture. Engineering New,s 38(26): 413.

Ausich, W.I. & Bottjer, D.J. 1982. Tiering in suspension-feeding communities on soft substrata throughout the Phanerozoic. Science, New Series 216(4542): 173–174.

Babcock, L.E. & Feldman, R.M. 1986. Devonian and Mississippian conulariids of North America, Part A. General description and *Conularia*. Annals of Carnegie Museum, 55: 349-410.

Bailey, J.B. 1983. Middle Devonian Bivalvia from the Solsville Member (Marcellus Formation), central New York State. American Museum of Natural History, Bulletin, 174: 193–325.

Bartlett, G.S. 1892. The cement industry. In: Milwaukee's great industries. Anderson, W.J. & Bleyer, J. (Eds.). Milwaukee, 352 pp.

Barton, E.E. 1886. Industrial history of Milwaukee. E.E. Barton publisher, Milwaukee, 222 pp.

Bather, F.A. 1908. The preparation and preservation of fossils. The Museums Journal, 7: 76–90.

Baumiller, T.K. & Gahn, F.J. 2002. Fossil record of parasitism on marine invertebrates with special emphasis on the platyceratid-crinoid interaction. Paleontological Society Special Papers, 8: 195–210.

Baxter, A.E. 1893. Hydraulic cement at Milwaukee. Stone 6(2): 108–110.

Beaver, H.H., Fay, R.O. & Moore, R.C. 1967. Glossary of Morphological Terms. In Moore, R.C, (Ed.), Treatise on Invertebrate Paleontology, Part S, Echinodermata 1, Volume 2, Blastoids. The Geological Society of America, Inc. & The University of Kansas. Boulder & Lawrence, pp. 345–350.

Bergland, M. & Hayes, P.G. 2014. Studying Wisconsin: The life of Increase A. Lapham. Wisconsin Historical Society Press, Madison, 410 pp.

Berthelet, W.T. 1942. The story of the Milwaukee Cement Company. Unpublished paper on file in the Milwaukee Cement Company papers, Wisconsin Historical Society, 9 pp.

Blodgett, R.B., Boucot, A.J. & Koch, W.F., II. 1988. New occurrences of color patterns in Devonian articulate brachiopods. Journal of Paleontology, 62: 46–51.

Boardman, R.S. & Cheetham, R.H. 1983. Glossary of Morphological Terms. In Robison, R.A., (Ed.), Treatise on Invertebrate Paleontology, Part G, Revised, Volume 1, Bryozoa. The Geological Society of America, Inc. & The University of Kansas. Boulder & Lawrence, pp. 304–321.

Boardman, R.S., Cheetham, R.H. & Cook, P.L. 1983. Introduction to the Bryozoa. In Robison, R.A., (Ed.), Treatise on Invertebrate Paleontology, Part G, Revised, Volume 1. The Geological Society of America, Inc. & The University of Kansas. Boulder & Lawrence, pp. 3–48.

Boyce, C.K., Hotton, C.L., Fogel, M.L., Cody, G.D., Hazen, R.M. & Knoll, A.H. 2007. Devonian landscape heterogeneity recorded by a giant fungus. Geology, 35: 399–402.

Boylan, J.C. 1973. *Eastmanosteus*, A placoderm from the Devonian of North America. PhD dissertation, Columbia University, 523 pp.

Bradley, J. 1973. *Zoophycos* and umbellule (Penatulacea): Their synthesis and identity. Palaeogeography, Palaeoclimatology, Palaeoecology, 13: 103–128.

Brett, C.E. 2003. Durophagous predation in Paleozoic marine benthic assemblages. Pp. 401–432 in: Kelley, P.H., Kowalewski, M., & Hansen, T.A. (Eds.). Predator-prey interactions in the fossil record. Kluwer Academic/Plenum Publishers, New York.

Brown, C.C. (Ed.) 1901. Directory of American cement industries and hand-book for cement users. Municipal Engineering Company, Indianapolis, Indiana, 645 pp.

Brown, W.L. 1987. Punctuated equilibrium excused: the original examples fail to support it. Biological Journal of the Linnean Society, 31: 383–404.

Burchard, E.F. 1912. Cement. Mineral resources of the United States for 1911, Part II-Nonmetals. United States Geological Survey, pp. 485–519.

Burton, C.J. & Eldredge, N. 1974. Two new subspecies of *Phacops? rana* (Trilobita) from the Middle Devonian of North-West Africa Palaeontology, 17: 349–363.

Carr, R.K. 1995. Placoderm diversity and evolution. Paris Museum of Natural History Bulletin, 17: 85–125.

Carr, R.K. & Jackson, G.L. 2009. The vertebrate fauna of the Cleveland Member (Famennian) of the Ohio Shale. Ohio Geological Survey Guidebook 22, Guide to the geology and paleontology of the Cleveland Member of the Ohio Shale.

Chamberlin, T.C. 1877. Devonian, Hamilton Cement Rock, Chapter IX, Geology of eastern Wisconsin. In: Geology of Wisconsin. Survey of 1873–1877, Volume II, Part 2, Published under the direction of the Chief Geologist by the Commissioners of Public Printing, Madison: 91–405.

Chamberlin, T.C. 1878. Devonian, Hamilton Cement Rock, Chapter IX, Geology of eastern Wisconsin. In: Geology of Wisconsin. Survey of 1873–1877, Volume II' Second Edition Revised, Part 2, Published under the direction of the Chief Geologist by the Commissioners of Public Printing, Madison: 91–405.

Chamberlin, T.C. 1883. Devonian Age, or Age of Fishes, Chapter X, General Geology, Volume 1, Part 1, In: Geology of Wisconsin. Survey of 1873–1879, Published under the direction of the Chief Geologist by the Commissioners of Public Printing, Madison: 201–212.

Cleland, H.F. 1907. Restorations of certain Devonian cephalopods with descriptions of new species. The Journal of Geology, 15: 459–469.

Cleland, H.F. 1911. The fossils and stratigraphy of the Middle Devonic of Wisconsin. Wisconsin Geological and Natural History Survey Bulletin 21 (Scientific Series No. 6): 222 pp.

Cooper, G.A. 1967. Age and correlation of the Tully and Cedar Valley Formations in the U.S. In: D.H. Oswald (Ed.) International Symposium on the Devonian System, vol. II: 701–709.

Cooper, G.A., et al. 1942. Correlation of the Devonian sedimentary formations of North America. Geological Society of America Bulletin, 53: 1729–1794.

Cordie, D.R. & Witzke, B.J. 2017. A new crinoid genus from the Middle Devonian of Iowa, USA (Camerata, Melocrinitidae). Paleontological Research, 2: 7–13.

Cox, L.R. 1960. Gastropoda. In Moore, R.C, (Ed.), Treatise on Invertebrate Paleontology, Part I, Mollusca 1. The Geological Society of America, Inc. & The University of Kansas. Boulder & Lawrence, pp. 84–169.

Cox, L.R. 1969. General Features of Bivalvia. In Moore, R.C., (Ed.), Treatise on Invertebrate Paleontology, Part N, Volume 1, Mollusca 6, Bivalvia. The Geological Society of America, Inc. & The University of Kansas. Boulder & Lawrence, pp. 3–121.

Day, J. 1998. Distribution of latest Givetian–Frasnian Atrypida (Brachiopoda) in central and western North America. Acta Palaeontologica Polonica, 43: 205–240.

Day, J. & Bunker, B.J. 1992. The stratigraphy, paleontology, depositional and diagenetic history of the Middle–Upper Devonian Cedar Valley Group of Central and Eastern Iowa. Iowa Department of Natural Resources Guidebook Series No. 6 (Prepared for North Central Section, Geological Society of America 26th Annual Meeting).

Dennis-Bryan, K. 1987. A new species of eastmanosteid arthrodire (Pisces: Placodermi) from Gogo, Western Australia. Zoological Journal of the Linnean Society, 90: 1–64.

De Rosa, R. 2001. Molecular data indicate the protostome affinity of brachiopods. Systematic Biology, 50: 848–859.

Donoghue, P.C.J, Forey, P.L. & Aldridge, R.J. 2000. Conodont affinity and chordate phylogeny. Biological Reviews (Cambridge), 75: 191–251.

Eastman, C.R. 1897. On the relation of certain plates in the dinichthyids. Bulletin of the Museum of Comparative Zoology, 31: 19–44.

Eastman, C.R. 1898a. Dentition of Devonian Ptyctodontidae (part 1 of 2). The American Naturalist, 32: 473–488.

Eastman, C.R. 1898b. Dentition of Devonian Ptyctodontidae (part 2 of 2). The American Naturalist, 32: 545–560.

Eastman, C.R. 1900. Dentition of some Devonian fishes. The Journal of Geology, 8: 32–41.

Eastman, C.R. 1907a. Devonic fishes of the New York formations. Memoir of the New York State Museum, 10: 235pp.

Eastman, C.R. 1907b. Fossil fishes in the collection of the United States National Museum. Proceedings of the U.S. National Museum, 52: 235–304.

Edwards, I. & Raasch, G.O. 1921. Notes on the occurrence of Upper Devonian strata in Wisconsin. Milwaukee Public Museum Yearbook for 1921, 1: 88–93.

Eldredege, N & Gould, S.J. 1972. Punctuated equilibria: An alternative to phyletic gradualism. Pp. 82–115 in: Schopf, A. (Ed.) Models in paleobiology. TJM Freeman, Cooper & Company, San Francisco.

Eldredge, N. 1972. Systematics and evolution of *Phacopss? rana* (Green, 1832) and *Phacops? iowensis* Delo, 1935 (Trilobita) from the Middle Devonian of North America. American Museum of Natural History, Bulletin, 147: 47–113.

Fay, R.O. & Wanner, J. 1967. Systematic Descriptions. In Moore, R.C, (Ed.), Treatise on Invertebrate Paleontology, Part S, Echinodermata 1, Volume 2, Blastoids. The Geological Society of America, Inc. & The University of Kansas. Boulder & Lawrence, pp. 396–454.

Fisher, D.W. 1962. Small Conoidal Shells of Uncertain Affinities. In Moore, R.C., (Ed.), Treatise on Invertebrate Paleontology, Part W, Miscellanea. The Geological Society of America, Inc. & The University of Kansas. Boulder & Lawrence, pp. 98–143.

Flower, R.H. & Gordon, M. 1972. New Nephriticeratidae (Nautilolidea) from the Devonian of Maryland, Pennsylvania, and Indiana. U.S. Geological Survey, Professional Paper 800-D, pp. D45–D50.

Ford, R.C. 2009. Conulariid test microstructure and mineralogy. Masters' thesis, Austin Peay State University.

Ford, R.C., Van Iten, H. & Clark, G.R. II. 2016. Microstructure and composition of the periderm of conulariids. Journal of Paleontology, 90: 389–399.

Gensel, P.G. & Andrews, H.N. 1984. Plant life in the Devonian. Praeger Publishers, New York.

Goldfuss, A. 1831. Petrefacta Germaniae. Arnz and Company, Düsseldorf.

Green, J. 1832. A monograph of the trilobites of North America: With colored models of the species. Clark & Raser, Philadelphia.

Green, 1838. Description of new trilobites. American Journal of Science (1st Series), 33: 406.

Greene, H. & Berthelet, W.T. 1949. The Milwaukee Cement Company. Wisconsin Magazine of History, 33(1): 28–39. Reprinted with numerous additional figures in Natural News by the Century House Historical Society.

Part I, pp. 9–10, Volume 8, Issue 2, Summer, 2006.

Part II, pp. 6–10, Volume 8, Issue 3, Fall, 2006.

Part III, conclusion, pp. 6–7, Volume 9, Issue 4, Winter, 2007–8.

Griesemer, A.D. 1965. Middle Devonian brachiopods of Southeastern Wisconsin. The Ohio Journal of Science, 65: 241–295.

Griffin, L.F. 1890. Natural and artificial cements. Popular Science Monthly, 37: 253–257.

Hall, J. 1843. Geology of New York. Part IV, Comprising the survey of the Fourth Geological District, Albany, 683 pp.

Hall, J. 1851. Upper Silurian and Devonian Series. In: J.W. Foster & J.D. Whitney, Report on the geology of the Lake Superior land district, Part II. U.S. 32nd Congress, Special Session, Senate Executive Document No. 4: 152–166.

Hall, J. 1861. Descriptions of new species of fossils; from the investigations of the Survey. In: Report of the superintendent exhibiting progress of the work, Hall, J. Madison, pp. 9–52.

Hall, J. 1862a. Physical geography and general geology. In: Hall, J. & Whitney, J.D., Report on the geological survey of the state of Wisconsin, Madison: 1–72.

Hall, J. 1862b. Palaeontology of Wisconsin. Report of the Geological Survey of Wisconsin, Volume 1: 54–448.

Hall, J. 1863. Contributions to paleontology, no. 6, On the occurrence of crustacean remains of the genera *Ceratiocaris* and *Dithryocaris*, with a notice of some new species from the Hamilton Group and Genesee Slate. Sixteenth Report of the New York State Cabinet of Natural History.

Hall, J. 1867. Containing the descriptions and figures of the fossil Brachiopoda of the Upper Helderberg, Hamilton, Portage and Chemung Groups. Paleontology of New York (4). C. Van Benthuysen, Albany.

Hayes, P.G. 1995. Increase A. Lapham: A useful and honored life. Wisconsin Academy Review, Spring, 1995: 10–15.
Henkel, E.G. 1938. Milwaukee Cement Works, 3 pp., and The Blue Hole, 4 pp., In: Shorewood History Project #7164, Vol. 1, on file in the Shorewood Public Library, Shorewood, Wisconsin.
Hickerson, W. 1997. Middle Devonian (Givetian) trilobite clusters from eastern Iowa and northwestern Illinois. Pp. 224–246 In: Brett, C.E. & Baird, G.C. (Eds.) Paleontological Events: Stratigraphic, ecological, and evolutionary implications. Columbia University Press, New York.
Hilderbrand, J.B & Zambito, J IV, 2015. Encrustation and taphonomic history of cephalopods from the Middle Devonian Milwaukee Formation. The Geological Society of America North-Central Section – 49th Annual Meeting, 2015. Paper No. 42–6.
Hill, D. 1981. Glossary of Morphological Terms Applied to Paleozoic Corals. In Teichert, C., (Ed.), Treatise on Invertebrate Paleontology, Part F, Supplement 1, Volume 1, Coelenterata, Rugosa and Tabulata. The Geological Society of America, Inc. & The University of Kansas. Boulder & Lawrence, pp. 32–36.
Hill, D. 1981a. Subclass Rugosa. In Teichert, C., (Ed.), Treatise on Invertebrate Paleontology, Part F, Supplement 1, Volume 1, Coelenterata, Rugosa and Tabulata. The Geological Society of America, Inc. & The University of Kansas. Boulder & Lawrence, pp. 1–762.
Hill, D. 1981b. Introduction to Tabulata. In Teichert, C., (Ed.), Treatise on Invertebrate Paleontology, Part F, Supplement 1, Volume 2, Coelenterata, Rugosa and Tabulata. The Geological Society of America, Inc. & The University of Kansas. Boulder & Lawrence, pp. 430–762.
Hoare, R.D. 1960. *Pentremitidea filosa* Whiteaves from the Silica Formation in Northwestern Ohio. Ohio Journal of Science, 60: 265–267.
Hore, R.E. 1915. Mineral resources of Michigan with statistical tables of production and value of Mineral products for 1914 and prior years. Publication 19, Geological Series 16, Michigan Geological and Botanical Survey.
Hueber, F.M. 2001. Rotted wood-alga-fungus: The history and life of *Prototaxites* Dawson, 1859. Review of Palaeobotany & Palynology, 116: 123–158.
Hussakof, L. & Bryant, W.L. 1918. Catalog of the fossil fish in the Museum of the Buffalo Society of Natural Sciences. Buffalo Society of Natural Sciences, Bulletin 12.
Jarochowska, F.H., Vinn, O. & Munnecke, A. 2016. Reducing taxonomic noise in problematic fossils: revision of the *incertae sedis* genus *Allonema* based on shape analysis. Bulletin of Geosciences, 91(1): 97–110.
Johnson, B.A. (Ed.). 2018. Women and Geology: Who are we, where have we come from, and where are we going? The Geological Society of America, Memoir 214.
Kidwell, S.M. & Jablonski, D. 1983. Taphonomic feedback ecological consequences of shell accumulation. Pp. 195–248 in: Tevesz, M.J.S. & McCall, P.L. (Eds.) Biotic interactions in Recent and fossil benthic communities. Springer Science & Business Media, New York.
Kloc, G.J. 2014. "*Phacops* vs. *Eldredgeops*." The Fossil Forum. www.thefossilforum.com/index.php?/topic/47202-phacops-vs-eldredgeops/&tab=comments#comment-509380
Kluessendorf, J. 1983. Observations on the commensalism of Silurian platyceratid gastropods and stalked echinoderms. Wisconsin Academy of Sciences, Arts, and Letters Bulletin, 71(1): 48–55.
Kluessendorf, J. & Mikulic, D.G. 1993. The Thomas A. Greene Memorial Museum and Collection. National Historic Landmark nomination, National Park Service.
Kluessendorf, J. & Mikulic, D.G. 1997a. Survey of scientifically and historically important bedrock geologic sites in southeastern Wisconsin. In: A regional natural areas study and critical species habitat protection and management plan for southeastern Wisconsin, Planning Report No. 42, Southeastern Wisconsin Regional Planning Commission, Waukesha, 531 pp.
Kluessendorf, J. & Mikulic, D.G. 1997b. Mr. Greene and his brachiopods. Rocks and Minerals, Vol. 72, January/February, pp. 50–55.
Kluessendorf, J. & Mikulic, D.G. 1998. Silurian and Devonian depositional environments of the westernmost Devonian rocks in the Michigan Basin. Geological Society of America Abstracts with Programs, 30(2): 28.
Kluessendorf, J. & Mikulic, D.G., 2008. The cement that made Milwaukee famous. Geological Society of America North-Central Section, Abstracts with Programs, v. 40(5): 65.
Kluessendorf, J., Mikulic, D.G. & Carman, M.R. 1988. Distribution and depositional environments of the westernmost Devonian rocks in the Michigan Basin. In: McMillan, N.J., Embry, A.F. & Glass, D.J. (Eds.). Devonian of the world. Canadian Society of Petroleum Geologists Memoir 14, Vol I: Regional synthesis: 251–263.
Klug, C.R. 1995. Conodonts from the Middle Devonian strata of southeastern Wisconsin. Geological Society of America Abstracts with Program 27(3): 65.
Klug, C.R. & Nelson, K.G. 1977. The Devonian of Wisconsin. 41st Annual Tri-State Geological Field Conference: A40–A46.
Koch, W.F. & Day, J.E. 1996. Late Eifelian–early Givetian (Middle Devonian) brachiopod paleobiogeography of eastern and central North America. In: P. Copper & Jin, J., Brachiopods. A.A. Balkema, Rotterdam: 135–143.
Land, J.E. 1883. Milwaukee, her trade, commerce and industries, 1882–3: manufacturing advantages, business and transportation facilities, together with sketches of the principle business houses and manufacturing concerns in the "Cream city," historical and descriptive review. Milwaukee.
Langson, W.J. 1876. Eighteenth annual report of the trade and commerce of Milwaukee for the year ending December 31, 1875 compiled for the Chamber of Commerce. The Sentinel Company, Milwaukee, 142 pp.
Langson, W.J. 1877. Nineteenth annual report on the trade and commerce of Milwaukee for the year ending December 31, 1876. Chamber of Commerce.
Langson, W.J. 1879. Twenty-First Annual Report on the Trade and Commerce of Milwaukee For the year ending December 31, 1878. Chamber of Commerce, Milwaukee, 164 pp.
Langson, W.J. 1891. Thirty-third annual report of the trade and commerce of Milwaukee for the year ending December 31st, 1890, and fiscal year of the Chamber of Commerce ending April 6th, 1891, compiled for the Chamber of Commerce. The Sentinel Company, Milwaukee, 217 pp.
Lapham, I.A. 1828. Notice of the Louisville and Shippingsport Canal, and of the geology of the vicinity. American Journal of Science

and Arts, 14(1): 65–69.
Lapham I.A. 1851. On the geology of the south-eastern portion of the State of Wisonsin; being the part not surveyed by the United States Geologists, in a letter to J.W. Foster. In: Report on the Geology of the Lake Superior Land District. J.W. Foster and J.D. Whitney, Part II, The Iron Region together with the General Geology, Chapter XI, Miscellaneous Observations on the Paleozoic Series. Washington, pp. 167–172.
Lapham, I.A. 1860. Discovery of Devonian Rocks and fossils in Wisconsin. American Journal of Science, Second Series, 29: 145.
Lapham. I.A. 1876. Geology. In: Atlas of the state of Wisconsin. H.F. Walling (Ed.), pp 16–19.
Leme, J.M. 2006. Análise cladística de Conulariidae Walcott (Neoproterozóico-Triassico): caracterizando e definindo um grupo de cnidários extintos. PhD dissertation, Universidade de São Paulo Instituto de Geociências.
Lewis, F.H. 1898. Plant of the Milwaukee Cement Company. The Engineering Record, 36(18): 382–384. Reprinted as Lewis, F.H., 1900. The plant of the Milwaukee Cement Company, Milwaukee, Wis. Chapter XVII. Pp. 169–177, in Lewis, F.H. (Ed.) The cement industry: Descriptions of Portland and natural cement plants in the United States and Europe, with notes on materials and processes in Portland cement manufacture. The Engineering Record, New York.
Lohr, J. 1964. Milwaukee's natural cement industry. Unpublished manuscript, 28 pp.
Long, J.A. 2011. The rise of fishes: 500 million years of evolution. The Johns Hopkins University Press, Baltimore.
Mayer, P.S. 1997. Stratigraphy and paleoecology of the Middle Devonian Berthelet Member of the Milwaukee Formation, Milwaukee County, Wisconsin. Geological Society of America Abstracts with Program, 29(4): 59.
Mayer, P.S. 2001. Biostratigraphy of the Milwaukee Formation and Antrim Shale (Middle to Late Devonian), Milwaukee, Wisconsin. Abstracts of the 2001 North American Paleontological Convention.
Mayer, P.S., Salkowski, N. & Turner, S. 1998. Late Devonian fish faunas from the Antrim Shale, Milwaukee County, Wisconsin. Geological Society of America Abstracts with Program, 29(4): 59.
McRoberts, C.A., Hegna, T.A., Burke, J.J., Stice, M.L., Mize, S.K. & Martin, M.J. 2013. Original spotted patterns on Middle Devonian phacopid trilobites from western and central New York. Geology, 41: 607–610.
Mikulic, D.G. 1979. The paleoecology of Silurian trilobites with a section on the Silurian stratigraphy of southeastern Wisconsin. Ph.D. dissertation, Oregon State University, Corvallis, 864 pp.
Mikulic, D.G. 1983. Milwaukee's gentlemen paleontologists. Wisconsin Academy of Arts, Sciences, and Letters Transactions, 71: 5–20. Reprinted in Rocks and Minerals, March/April, 1991, 66: 136–146.
Mikulic, D.G. 1991. The Greene Geological Museum: Science in the 19th century and today. National Park Service, Cultural Resource Management, 14: 6–7.
Mikulic, D.G., & Kluessendorf, J. 1982. Preservation of scientifically and historically important geologic sites in Milwaukee County, Wisconsin. Southeastern Wisconsin Regional Planning Commission Technical Record, 4: 1–13.
Mikulic, D.G. & Kluessendorf, J. 1988a. Subsurface stratigraphic relationships of the Upper Silurian and Devonian rock of Milwaukee County, Wisconsin. Geoscience Wisconsin, 12: 1–23.
Mikulic, D.G. & Kluessendorf, J. 1988b. Geologic logs from Milwaukee County, Wisconsin. Wisconsin Geological and Natural History Survey Open-file Report WOFR 88-1, 7 pp.
Mikulic, D.G. & Kluessendorf, J. 1998. Sequence stratigraphy and depositional environments of the Silurian and Devonian rocks of southeastern Wisconsin. SEPM Great Lakes Section/Michigan Basin Geological Society Fall Field Conference guidebook.
Mikulic, D.G. & Kluessendorf, K. 2001. Gilbert O. Raasch, student of Wisconsin's ancient past. Geoscience Wisconsin, 18: 75–93.
Miller, M.F., Cowan, E.A. & Nielsen, H.H. 2009. Significance of the trace fossil *Zoophycos* in Pliocene deposits, Antarctic continental margin (ANDRILL 1B drill core). Antarctic Science, 21: 609–618.
Milwaukee Cement Company. 1897. Views of the cement co.'s works. Milwaukee, Wisconsin.
Monroe, C.E. & Teller, E.E. 1899. The fauna of the Devonian formation at Milwaukee, Wisconsin. The Journal of Geology, 7: 272–283.
Moore, R.C. 1978. Glossary of Crinoid Morphological Terms. In Moore, R.C. & Teichert, C, (Eds.), Treatise on Invertebrate Paleontology, Part T, Echinodermata 2, Volume 1. The Geological Society of America, Inc. & The University of Kansas. Boulder & Lawrence, pp. 229–242.
Moysiuk, J., Smith, M.R. & Caron, J-B. 2017. Hyoliths are Palaeozoic lophophorates. Nature, 541: 394–397.
Nagel-Meyers, J., Dietl, G. & Brett, C. 2011. Ecological stasis in the Middle Devonian Hamilton fauna: How stable are species interactions? 2011 GSA Annual Meeting in Minneapolis, Geological Society of America Abstracts with Programs, 43(5): 30.
National Park Service. 2016. National historic landmarks program list of national historic landmarks by state. U.S. Department of the Interior, National Park Service.
Nelson, J.S. 1994. Fishes of the World (3rd ed.). John Wiley & Sons, Inc., New York.
Nelson, J.S., Grande, T.C. & Wilson, M.V.H. 2016. Fishes of the world (5th ed.). John Wiley & Sons, Inc., Hoboken.
Nelson, K.G. & Roberts, J. 1963. The Milwaukee Formation along Lake Michigan's shore. Wisconsin Academy of Sciences, Arts and Letters, 52: 77–81.
Newberry, J.S. 1889. The Paleozoic fishes of North America. Monograph of the United States Geological Survey, 16. 340 pp.
Peckham, S.F. 1901. Cement for a modern street. Popular Science Monthly, 60: 145–157.
Pemberton, S.G. & Frey, R.W. 1982. Trace fossil nomenclature and the *Planolites-Palaeophycus* dilemma. Journal of Paleontology, 56: 843–881.
Penhallow, D.P. 1896. *Nematophyton ortoni*, n. sp. Annals of Botany, 10: 41–48.
Penhallow, D.P. 1908. Fossil plants from the Middle Devonian of Milwaukee, Wisconsin. Wisconsin Natural History Society Bulletin, 6: 8–12.
Platt, L.A. 2010. Planning ideology and geographic thought in the early twentieth century: Charles Whitnall's Progressive Era Park designs for Socialist Milwaukee. Journal of Urban History, 36(6): 771–791.
Pohl, E.R. 1929. Middle Devonian pelecypods of Wisconsin and their bearing on correlation. Journal of the Washington Academy of Science, 19: 53–59.
Pohl, E.R. 1929. The Devonian of Wisconsin Part 1, Lamellibranchiata. Bulletin of the Public Museum of the City of Milwaukee, 11, 100 pp.
Pojeta, J. Jr., Renjie, Z. & Zunyi, Y. 1986. Systematic palaeontology of Devonian pelecypods of Guangxi and Michigan. Pp. 57–108

in: Pojeta, J. Jr. (Ed.) Devonian rocks and Lower and Middle Devonian pelecypods of Guangxi, China, and the Traverse Group of Michigan. U.S. Geological Survey Professional Paper 1394.

Preslicka, J.E., Newson, C.R. & Blume, T.E. 2013. Fauna from the Brooks Hardground Bed: Lower Solon Member, Little Cedar Formation, Bruening Inc. Brooks Quarry, Independence, Iowa. Pp. 41–51 in: Day, J., Witzke, B. & Lundy, S. (Eds.) Aspects of the Paleozoic history of epeiric seas of the Iowa Basin. Iowa Geological and Water Survey Guidebook Series No. 29.

Raasch, G.O. 1935. Devonian of Wisconsin. Kansas Geological Society Guidebook for the Annual Field Conference no. 9, pp. 261–267.

Reimann, I.G. 1945. New Devonian blastoids. Bulletin of the Buffalo Society of Natural Sciences, 19(2): 22–41.

Roberts, J.E. 1972. Agglutinated foraminifera from the Devonian of Wisconsin. Master's thesis, University of Wisconsin-Milwaukee.

Rolfe, W.D.I. 1969. Phyllocarida. In Moore, R.C. (Ed.), Treatise on Invertebrate Paleontology, Part R, Arthropoda 4, Volume 1. The Geological Society of America, Inc. & The University of Kansas Press. Boulder & Lawrence, pp. 296–328.

Rovey, C.W., II. 1990. Stratigraphy and sedimentology of Silurian and Devonian carbonates, eastern Wisconsin, with implications for ground-water discharge into Lake Michigan. Ph.D. dissertation, University of Wisconsin-Milwaukee, 427 pp.

Rovey, C.W., II. 1997. Proposed reference sections and correlation of Upper Silurian and Devonian strata, eastern Wisconsin. Geoscience Wisconsin, 16: 37–46.

Schoenemann, B., Clarkson, E.N.K. & Ryck, U. 2014. Colour patterns in Devonian trilobites. The Open Geology Journal, 8: 113–117.

Schumacher, D. 1967. Conodonts and biostratigraphy of the Middle and Upper Devonian of Wisconsin. Master's thesis, University of Wisconsin.

Schumacher, D. 1971a. Conodonts and biostratigraphy of the "Kenwood" Shale (Upper Devonian). In: D.L. Clark (Ed.) Conodonts and biostratigraphy of the Wisconsin Paleozoic. Wisconsin Geological and Natural History Survey Information Circular, 19: 68–77.

Schumacher, D. 1971b. Conodonts from the Middle Devonian Lake Church and Milwaukee Formations. In: D.L. Clark (Ed.) Conodonts and biostratigraphy of the Wisconsin Paleozoic. Wisconsin Geological and Natural History Survey Information Circular, 19: 55–67.

Schumacher, D. 1971c. Lake Church and Milwaukee conodonts. In: D.L. Clark (Ed.) Conodonts and biostratigraphy of the Wisconsin Paleozoic. Wisconsin Geological and Natural History Survey Information Circular, 19: 90–99.

Sessa, E.B. 2018. Evolution and classification of ferns and lycophytes. Pp. 179–200 in: Fernandez, H. (Ed.) Current advances in fern research. Springer, New York.

Shorewood Historical Society, 2016. Milwaukee Cement Company Mill #1. Shorewood, Issue 4. http://shorewoodhistory.org/newsletters/ShorHistNews10-16.pdf Downloaded 01-Jun-2018.

Signor, P.W., III & Brett, C. 1984. The Mid-Paleozoic precursor to the Mesozoic marine revolution. Paleobiology, 10: 229–245.

Springer, F. 1920. The Crinoidea Flexibilia (With an atlas of A., B., C., and 76 plates). Smithsonian Institution Publication, 2501, Part 2.

Sprinkle, J. 1973. Morphology and evolution of blastozoan echinoderms. Harvard University, Museum of Comparative Zoology Special Publication, 283 pp.

Stickney, G.F. Great industries of Wisconsin X Miscellaneous. The Milwaukee Sentinel (Milwaukee, WI), Sunday, 1897 (12 Dec.), Issue 53: 23.

Struve, X. 1990. Palaeozoologie III (1896-1990). Courier Forschunginstitut Senkenberg, 127: 251–279

Stumm, E.C. 1953. Trilobites of the Devonian Traverse Group of Michigan. University of Michigan, Contributions from the Museum of Palaeontology, 10: 101–157.

Stumm, E.C. 1967. *Planalveolitella*, a new genus of Devonian tabulate corals, with a redescription of *Planalveolites* (Edwards and Haime). The University of Michigan, Contributions from the Museum of Paleontology, 21: 67–72.

Sumpter, P.M. & Meyer, J.J. 1991. Catalog of fossil type and figured specimens in the Milwaukee Public Museum. Milwaukee Public Museum, Contributions in Biology and Geology, Number 82.

Sweet, W.C. 1964. Cephalopoda—General Features. In Moore, R.C. (Ed.), Treatise on Invertebrate Paleontology, Part K, Mollusca 3, Cephalopoda-General Features-Endoceratoidea-Actinoceratoidea-Nautiloidea-Bactritoidea. The Geological Society of America, Inc. & The University of Kansas Press. Boulder & Lawrence, pp. 4–13.

Taylor, P.D. & Vinn, O. 2006. Convergent morphology in small spiral worm tubes ('*Spirorbis*') and its palaeoenvironmental implications. Journal of the Geological Society, London, 163: 225–228.

Taylor, P.D. & Wilson, M.A. 2008. Morphology and affinities of Hederelloid "bryozoans." Pp. 301–309 in: Hageman, S.J., Key Jr., M.M. & Winston, J.E. (Eds.) Bryozoan studies 2007 Proceedings of the 14th International Bryozoology Association Conference, Boone, North Carolina, July 1–8, 2007. Virginia Museum of Natural History, Martinsville, VA.

Teichert, C. 1964. Glossary of Morphological Terms Used for Nautiloids. In Moore, R.C. (Ed.), Treatise on Invertebrate Paleontology, Part K, Mollusca 3, Cephalopoda-General Features-Endoceratoidea-Actinoceratoidea-Nautiloidea-Bactritoidea. The Geological Society of America, Inc. & The University of Kansas Press. Boulder & Lawrence, pp. 54–59.

Teller, E.E. 1900. The Hamilton Formation at Milwaukee, Wisconsin. Wisconsin Natural History Society Bulletin, 1: 47–56.

Thwaites, R.G. (Ed.) 1900. Travels and Explorations of the Jesuit Missionaries in New France. Vol. LX, Lower Canada, Illinois, Iroquois, Ottawas: 1675–1677. Cleveland, 323 pp.

Toepfer, P. 1977. The Blue Hole of the Milwaukee River. Wisconsin Natural Resources, 1:3, pp. 4–5.

Turner, S., Burrow, C.J., Schultze, H.P., Blieck, A., Reif, W.E., Rexroad, C.B. & Nowlan, G.S. 2010. False teeth: Conodont-vertebrate phylogenetic relationships revisited. Geodiversitas, 32: 545–594.

Ubaghs, G. 1967. General Characters of Echinodermata. In Moore, R.C. (Ed.), Treatise on Invertebrate Paleontology, Part S, Echinodermata 1. The Geological Society of America, Inc. & The University of Kansas Press. Boulder & Lawrence, pp. 3–60.

Ubaghs, G., Breimer, A., Lane, N.G., Wienberg Rasmussen, H., Moore, R.C., Brower, J.C., Sieverts-Doreck, H., Strimple, H.L., Macurda, Jr., D.B., Meyer, D.L. & Roux, M. 1978. General Features of Crinoidea. In Moore, R.C. & Teichert, C. (Eds.), Treatise on Invertebrate Paleontology, Part T, Echinodermata 2, Volume 1. The Geological Society of America, Inc. & The University of Kansas Press. Boulder & Lawrence, pp. 9–347.

Van Iten, H., Leme, J.M., Rodrigues, S.C. & Simões, M.G. 2006a. New data on the anatomy of *Conularia milwaukeensis* Cleland, 1911 (Middle Devonian, Iowa and Wisconsin). Journal of Paleontology, 80: 392–395.

Van Iten, H., Leme, J.M., Simões, M.G, Marques & Collins, A.G. 2006b. Reassessment of the phylogenetic position of conulariids in the subphylum Medusozoa (phylum Cnidaria). Journal of Systematic Palaeontology, 4: 109–118.

Van Iten, H., Leme, J. M., Pacheco, M.L.A.F., Simões, M.G., Fairchild, T.R., Rodrigues, F., Galante, D., Boggiani, P.C. & Marques, A.C. 2016a. Origin and early diversification of phylum Cnidaria: key macrofossils from the Ediacaran System of North and South America. Pp. 31–40 in: Goffredo, S. & Dubinsky, Z. (Eds.), The Cnidaria, Past, Present and Future. Springer International Publishing, Switzerland.

Van Iten, H., Marques, A.C., Leme, J.M., & Simões, M.G. 2014. Origin and early evolution of the phylum Cnidaria Verrill: major developments in the analysis of the taxon's Proterozoic–Cambrian history. Palaeontology, 3: 1–14.

Van Iten, H., Muir, L., Simões, M.G, Leme, J.M., Marques, A.C. & Yoder, N. 2016b. Palaeobiogeography, palaeoecology and evolution of Lower Ordovician conulariids and Sphen othallus (Medusozoa, Cnidaria), with emphasis on the Fezouata Shale of southeastern Morocco. Palaeogeography, Palaeoclimatology, Palaeoecology, 460: 170–178.

Van Iten, H. & Südkamp, W. 2010. Exceptionally preserved conulariids and an edrioasteroid from the Hunsrück Slate (Lower Devonian, SW Germany). Palaeontology, 53: 403–414.

Van Iten, H., Tollerton, V.P., Jr., Ver Straeten, C.A., Leme, J.M., Simões, M.G. & Rodrigues, S.C. 2012. Life mode of in situ Conularia in a Middle Devonian epibole. Palaeontology, 56: 1–20.

Vinn, O. & Mutvei, H. 2009. Calcareous tubeworms of the Phanerozoic. Estonian Journal of Earth Sciences, 58: 286–296.

Vinn, O. & Zaton, M. 2012. Phenetic phylogenetics of tentaculitoids – extinct, problematic calcareous tube-forming organisms. GFF, 134: 145–156.

Weller, S. 1898. Descriptions of Devonian crinoids and blastoids from Milwaukee, Wisconsin. Annals of the New York Academy of Sciences, 11: 117–126.

Whitfield, R.P. 1882. Species from the Hamilton Group. In: Geology of Wisconsin. Survey of 1873–1879, Volume IV, Part III, Palaeontology. Published under the direction of the Chief Geologist by the Commissioners of Public Printing, Madison, pp. 167 and 324–340.

Whittington, H.B. & Kelly, S.R.A. 1997. Morphological Terms Applies to Trilobita. In Kaesler, R.L. (Ed.), Treatise on Invertebrate Paleontology, Part O, Arthropoda 1 Trilobita, Revised, Volume 1, Introduction, Order Agnostida, Order Redlichiida. The Geological Society of America, Inc. & The University of Kansas. Boulder & Lawrence, pp. 313–329.

Whittington, H.B., Chang, W.T., Doan, W.T., Fortey, R.A., Jell, P.A., Lauric, J.R., Palmer, A.R., Repina, L.N., Rushton, A.W.A. & Shergold, J.H. 1997. Systematic Descriptions of the Class Trilobita. In Kaesler, R.L. (Ed.), Treatise on Invertebrate Paleontology, Part O, Arthropoda 1 Trilobita, Revised, Volume 1, Introduction, Order Agnostida, Order Redlichiida. The Geological Society of America, Inc. & The University of Kansas. Boulder & Lawrence, pp. 331–530.

Whittemore, D.J., 1875. Hydraulic Cement. Paper read before the Fortnightly Club of Milwaukee, 11 pp.

Williams, A. & Brunton, H.C. 1997. Morphological and Anatomical Terms Applied to Brachiopods. In Kaesler, R.L. (Ed.), Treatise on Invertebrate Paleontology, Part H, Revised, Volume 1, Brachiopoda, Introduction. The Geological Society of America, Inc. & The University of Kansas. Boulder & Lawrence, pp.423–440.

Williams, A., Carlson, S.J. & Brunton, H.C. 2000a. Phylum Brachiopoda. In Kaesler, R.L., (Ed.), Treatise on Invertebrate Paleontology, Part H, Revised, Volume 2, Brachiopoda, Linguliformea, Craniiformea, and Rhynchonelliformea. The Geological Society of America, Inc. & The University of Kansas. Boulder & Lawrence, pp. 28–29.

Williams, A., Carlson, S.J. & Brunton, H.C. 2000c. Subphylum Craniiformea. In Kaesler, R.L. (Ed.), Treatise on Invertebrate Paleontology, Part H, Revised, Volume 2, Brachiopoda, Linguliformea, Craniiformea, and Rhynchonelliformea. The Geological Society of America, Inc. & The University of Kansas. Boulder & Lawrence, p. 158.

Williams, A., Carlson, S.J. & Brunton, H.C. 2000b. Subphylum Linguliformea. In Kaesler, R.L. (Ed.), Treatise on Invertebrate Paleontology, Part H, Revised, Volume 2, Brachiopoda, Linguliformea, Craniiformea, and Rhynchonelliformea. The Geological Society of America, Inc. & The University of Kansas. Boulder & Lawrence, p. 30.

Williams, A., Carlson, S.J. & Brunton, H.C. 2000d. Subphylum Rhynchonelliformea. In Kaesler, R.L. (Ed.), Treatise on Invertebrate Paleontology, Part H, Revised, Volume 2, Brachiopoda, Linguliformea, Craniiformea, and Rhynchonelliformea. The Geological Society of America, Inc. & The University of Kansas. Boulder & Lawrence, p. 193.

Wilson, M.A. & Taylor, P.D. 2014. The morphology and affinities of *Allonema* and *Ascodictyon*, two abundant Palaeozoic encrusters commonly misattributed to the ctenostome bryozoans. In: Rosso, A., Wyse Jackson, P.N. & Porter, J. (Eds.) Bryozoan studies 2013. Studi trentini di scienze naturali, 94: 259–266.

Witzke, B.J. & Bunker, B.J. 1997. Sedimentation and stratigraphic architecture of a Middle Devonian (late Givetian) transgressive-regressive carbonate-evaporite cycle, Coralville Formation, Iowa area. Pp. 67–88 in: Klapper, G., Murphy, M.A. & Talent, J.A. (Eds.), Paleozoic sequence stratigraphy, biostratigraphy, and biogeography: Studies in honor of J. Granville ("Jess") Johnson. Geological Society of America Special Paper 321.

Woodward, A.S. 1891. Catalogue of the fossil fishes in the British Museum (Natural History), Part 2. British Museum of Natural History, London.

Yonge, C.M. 1960. General Characters of the Mollusca. In Moore, R.C. (Ed.), Treatise on Invertebrate Paleontology, Part I, Mollusca 1. The Geological Society of America, Inc. & The University of Kansas. Boulder & Lawrence, p. 13–136.

Zambito, J. & Day, J. 2015. Chronostratigraphic revision of the Givetian (Middle Devonian) Milwaukee Formation. Geological Society of America North-Central Section – 49th Annual Meeting, Paper No. 10.

Zaton M. & Vinn, O. 2011. Microconchids and the rise of modern encrusting communities. Lethaia, 44: 5–7.

Ziemke, P.C. 1935. Milwaukee blasts a river bed. The Explosives Engineer, 13:1, pp. 7–10.

Zhu, M., Yu, X., Ahlberg, P.E., Choo, B., Lu, J., Qiao, T., Qu, Q., Zhao, W., Jia, L., Blom, H. & Zhu, Y. 2013. A Silurian placoderm with osteichthyan-like marginal jaw bones. Nature, 502: 188–193.

Glossary

Abaxial—Directed away from the longitudinal axis of an organism or structure.
Acanthodii—The class of teleosts having dermal and perichondral bone as well as a series of stout spines before its dorsal, anal and paired fins. This class was, for a short time, included within the Placodermi, but is now recognized to have more features in common with the Osteichthyes.
Actinoceratoidea—The subclass of cephalopods having shells of medium to large size ranging in shape from straight and slender to slightly curved, simple, curved septa and sutures, and typically a siphuncle that was broadly expanded between septa and at least partially filled with calcareous deposits.
Abaxial—Directed toward the longitudinal axis of an organism or structure.
Ambulacrum (pl. a)—Structure in echinoderms (generally five or multiples of five in number) in which food is conveyed to the mouth.
Ammonoidea—An extinct subclass of marine cephalopods bearing an external, chambered shell divided by partitions (septa) that have convoluted margins which are often sufficiently distinct to aid in identification of species.
Amphibian—Member of the Amphibia, a class of cold-blooded vertebrates having gilled aquatic larvae and air-breathing adults; includes frogs, toads, salamanders and newts.
Animalia (=Metazoa)—The kingdom of multicellular eukaryotes that typically feed on organic matter, process it in a central cavity, and are mobile for at least part of their lives.
Annelida—The protostome animal phylum of aquatic or terrestrial, vermiform, bilaterally symmetrical, coelomate metazoans that have a segmented external cuticle that is never molted; includes earthworms, leeches, marine worms and many other worn-like animals.
Anthozoa—A class of aquatic (mostly marine) polypoid cnidarians having a tubular body, tentacles around their mouths and generally sedentary adults. Unlike the other cnidarian classes, they do not have a medusa stage in their development. Includes corals, sea anemones, and sea fans.
Antrim Shale—An Upper Devonian formation lying unconformably on the Milwaukee Formation; known in Wisconsin only from glacial erratic blocks, float from tunnel excavations, and core samples; the Wisconsin material is the western-most occurrence of the Antrim Shale, which is found more extensively in Ohio, Indiana and the type locality, eastern Michigan.
Apatite—A group of calcium phosphate minerals whose members are major constituents of bones, teeth and certain shells and other skeletal materials of such invertebrates as lingulid brachiopods and the extinct conodonts and conulariids.
Aragonite—An unstable morph of calcium carbonate, generally having a pearly luster, found primarily in certain layers of the shells of brachiopods and mollusks; transforms into calcite over time.
Arborescent—Tree-like in form and size.
Archaebacteria (=Archaea)—The domain of single celled organisms that are essentially intermediate between the two other domains, Eubacteria and Eucaryota; includes forms that can live in extreme conditions such as hydrothermal vents, hot springs, and hypersaline water.
Argillaceous—Pertaining to rocks composed of grains less than 0.0625 mm in diameter, i.e., clay rich.
Arthropoda—The protostome animal phylum having a chitinous exoskeleton and typically segmented body with jointed legs; includes the chelicerates, trilobites, myriapods, crustaceans, insects, and many other living and extinct aquatic and terrestrial groups; the most diverse phylum of animals.
Articulated—Hinged or otherwise held together, generally via interlocking parts.
Asphaltum—A black, tar-like substance occurring in rock cavities and derived from life forms through natural processes.
Asterozoa—A more-or-less star-shaped class of echinoderms typically having five (or multiples of five) appendages, each with a series of tube feet (hydraulically operated by a water vascular system) on either side of the ambulacra; includes sea stars (starfish) and brittle stars.
Benthic—The marine zone including the region on or beneath the sea floor.
Berthelet Member—The lowest (oldest) member of the Milwaukee Formation, overlying the Thiensville Formation and underlying the Lindwurm Member; comprised of medium to thick bedded argillaceous limestone and argillaceous dolomite; fossils generally in the form of internal and external molds; the source of most of the Milwaukee Formation's fossil fish and land plants.
Biocalcarenite—A detrital carbonate rock composed of sand-sized particles largely derived from animal and/or protist shells.
Biota—The animals, plants, fungi and other life forms of a particular place, time, environment, or geological unit.
Bioturbation—The restructuring of sedimentary deposits by moving organisms.
Bivalvia—The class of aquatic mollusks that includes the clams, oysters, mussels and related forms typically possessing two calcareous shells that hinge together and enclose a muscular foot and other soft parts.
Blastoidea—An extinct class of echinoderms with biserial brachioles bordering their ambulacral areas, well developed pentameral symmetry, seventeen major plates (basals, radials and deltoids) arranged in three circlets, a lancet plate underlying its ambulacral system, an anus, mouth, hydrospires on either side of each ambulacrum crossing the plate suture, and typically a spiracle near the adoral extremity of each ambulacrum.
Blastopore—The first opening formed in the central cavity of an embryo during gastrulation.
Brachial valve (=dorsal valve)—Valve that invariably contains any skeletal support for lophophore and never wholly accommodates pedicle; commonly smaller than ventral valve and with distinctive muscle-scar pattern.
Brachidium (pl. a)—One of the pair of calcareous supports for the lophophore of brachiopods, in the form of a loop or spire.
Brachium (pl. a) (=arm)—In certain crinozoans (e.g., crinoids), one of the food-gathering appendages attached to the top of the calyx.
Brachiole—In blastoids, one of the slender food-gathering appendages attached to the lancet plates at the border of the ambulacra.
Brachiopoda—The lophophorate animal phylum consisting of solitary, marine, sessile, coelomate, metazoans having a pair of calcareous or chitinophosphatic valves that are bilaterally symmetrical about the median plane that is normal to the surface of separation between valves.
Breccia—Sedimentary rock composed primarily of large, angular or subangular fragments cemented by a matrix of smaller particles that may be of a different composition.
Bryozoa—The lophophorate animal phylum consisting of colonial, aquatic, generally sessile, polymorphic metazoans, regarded as coelomate, with a retractable food-gathering lophophore, U-shaped digestive tract and typically having a calcareous skeleton.

Calcareous—Composed largely or entirely of at least one morph of calcium carbonate.
Calcining—The heating of solids to a high temperature for the purpose of removing volatile substances, oxidizing a portion of mass, or rendering them friable.
Calcite—The most common morph of calcium carbonate; the primary mineral of which limestone and many calcareous shells are made.
Calyx (=Cup) (Zoological context)—The head region of a crinozoan exclusive of its brachia/brachioles, enclosing the main portion of the visceral mass.
Cambrian—The first period in the Paleozoic Era, ranging from about 541,000,000 to 485,000,000 years ago, from which the oldest abundant fossils are encountered; followed by the Ordovician Period.
Carboniferous—The period in the Paleozoic Era between the Devonian and Permian Periods, ranging from about 359,000,000 to 299,000,000 years ago.
Carina (pl. ae)—In conulariids, an internal, thickened ridge located along the corners and/or midlines of the periderm.
Caustic potash—Potassium hydroxide (KOH); a strong alkali once used to remove calcareous matrix from fossils; generally fallen into disuse due to its health hazards and destructive effects on the fossils themselves.
Cenozoic—The era following the Mesozoic Era, including the Paleogene, Neogene and Quaternary periods; last 65 million years of Earth history.
Cephalon (pl. a) (Zoological context)—The head region of certain arthropods, e.g., trilobites.
Cephalopoda—The class of agile, bilaterally symmetrical, predacious molluscan carnivores typically having a pair of complex eyes, a beak, a set of eight or more elongated, flexible appendages, and a funnel for propulsion through the water; have either a single calcareous external or internal shell or may lack a shell entirely.
Chaetognatha (=Arrow worm)—The phylum of planktonic and nektonic, coelomate, bilaterally symmetrical, mostly marine and predatory, wormlike animals.
Chert—An opaque sedimentary rock composed of extremely fine-grained crystals of silica.
Chimaeroidei—The chondrichthyan suborder that typically has grinding tooth plates and a prominent leading spine on the first dorsal fin; includes the ratfish.
Chitin (Biochemical context)—A stiffening yet flexible nitrogenous material of which fingernails, hooves and many invertebrate exoskeletons (especially arthropods) are primarily composed.
Chitinophosphatic—Composed of chitin and phosphates (especially calcium phosphate), such as the outer covering of conulariids and the shell of lingulid brachiopods.
Chitinozoan—Enigmatic, flask-shaped marine microorganisms or egg cases (Ordovician–Holocene) typically composed of chitin-like material.
Chondrichthyes—The class of gnathostomes having cartilagenous endoskeletons—the males having a pair of claspers; includes the sharks, rays, chimaeras, and related forms.
Chordata—The phylum of bilaterally symmetrical animals having a notochord and pharyngial gill slits in adults or certain embryological stages; includes the urochordates (tunicates), cephalochordates (lancelets), craniates, and, tentatively, conodonts.
Claspers—The paired organs of the male of certain animals, such as chondrichthyans and certain placoderms, used for stabilization during mating.
Class—The rank in taxonomic hierarchy between phylum and order.
Cnidaria—The phylum of acoelomate, headless, aquatic, predatory, colonial or solitary, polypoid or medusoid animals having a mouth surrounded by stinging-cell-bearing tentacles; includes corals, sea fans, hydroids, and jellyfish.
Coelacanthiformes—The sarcopterygian order that includes the extant *Latimeria*, classically labeled a "living fossil."
Coelom—A fluid-filled body cavity lined with mesodermal cells; found in vertebrates and many other bilaterally symmetrical animals.
Coleoidea—The subclass of cephalopods that includes most of the living forms, such as octopi, squids, and cuttlefish; typically shelless, but often with a hard internal structure, such as a gladius (in many squids), cuttlebone (in cuttlefish) or thickened rostrum (in the extinct belemnoids).
Column—In most stalked echinoderms, an elongated cylindrical structure on the aboral end of the calyx with which it attaches to the substrate at its other extremity; sometimes referred to as a stalk.
Columnal—The unit ossicle composing the crinozoan column; typically discoidal in shape.
Commensalism—A relationship between individuals of two species (A and B) in which A benefits from B without either harming or benefiting B.
Commensal—The species that benefits from the commensalistic relationship.
Conodonta—The class of small, marine, pelagic and benthic chordates of eel-like form having head regions containing assemblages of microscopic phosphatic tooth-like structures also termed conodonts.
Conulariida—The extinct group of sessile benthic, polypoid scyphozoan cnidarians having a steeply pyramidal, generally four-sided periderm constructed of alternating, extremely thin organic and apatitic lamellae that in some species could be folded to cover the oral end of the animal.
Corallite—The exoskeleton of a solitary polyp or of a bud in a colony of anthozoans.
Craniata—The chordate subphylum characterized by possession of a cranium.
Cranidium—In trilobites, the central portion of the exoskeleton of the cephalon bounded laterally by the facial suture.
Cretaceous—A period in the Mesozoic Era between the Jurassic and Tertiary, ranging from about 145,000,000 to 66,000,000 years ago.
Crinoid—Member of the Crinoidea, a class of stalked echinoderms with pentamerous symmetry and five (or multiples of five) outspread, branched, food-gathering arms (brachia).
Crinozoa—The subphylum of echinoderms typically attached by a stalk and possessing a calyx having ambulacra that served as food grooves extending outward onto projecting appendages (brachia or brachioles).
Crustacea—The subphylum of mandibulate arthropods typically having two pairs of pre-oral appendages; includes crabs, lobsters, isopods (wood lice), amphipods, copepods, ostracods and other living and extinct forms.
Ctenophora—The phylum of carnivorous, marine, generally free swimming animals placed until recently in the same phylum with the Cnidarians; unlike cnidarians, ctenophores lack stinging cells and bear few to no tentacles, having instead a series of ciliated combs for locomotion and a sticky substance for capturing prey; comb jellies.

Cyanobacteria (=Blue green "algae")—The phylum of photosynthetic eubacteria accounting for much Earth's oxygen; some types can form multicellular colonies of biofilms, biomats or stromatolites, in some cases working in symbiosis or metabolic cooperation with algae and other types of organisms.
Dendroidea—The order of sessile graptolites attached by the apex of the sicula, which is then generally more or less embedded in secondary cortical tissue forming a root-like base.
Dermal plate—One of the bony plates hinged or joined together in the head and thoracic regions of placoderm fishes forming a body armor either on the surface or beneath a layer of skin.
Deuterostomia—The subkingdom of animals having a mouth that delelops from a secondary opening in the embryonic gut other than the blastopore; includes the Bryozoa, Brachiopods, Chaetognatha, Echinodermata, Hemichordata, and Chordata, and several minor phyla.
Detritus—Loose material, such as rock fragments or organic particles, that results directly from disintegration rather than from crystallization or other means.
Detrivore—One who feeds on detritus.
Devonian—The period in the Paleozoic Era between the Silurian and Carboniferous Periods, ranging from about 419,000,000 to 359,000,000 years ago; the period in which the Milwaukee Formation was deposited.
Disarticulated—Unhinged or otherwise disjoined.
Disconformity—An unconformity such that the beds above and below the surface are parallel.
Dissolution (Geological context)—The breakdown of a material into individual ions, atoms or molecules by water or other solvents.
Dolomite—The mineral calcium magnesium carbonate, generally including lesser amounts of iron in place of the calcium and magnesium.
Dolomitization—The process by which limestone (calcium carbonate) transforms to dolomite via replacement of some of the calcium ions by magnesium ions.
Domain—Generally regarded as the highest rank in taxonomic hierarchy, positioned immediately above kingdom.
Dorsal valve (=brachial valve)—Valve that invariably contains any skeletal support for lophophore and never wholly accommodates pedicle; commonly smaller than ventral valve and with distinctive muscle-scar pattern.
Doublure—In trilobites, the reflexed continuation of the dorsal exoskeleton onto the ventral surface.
Dysaerobic—Pertaining to a depositional environment having 0.1 to 1.0 ml of dissolved oxygen per liter of water.
Echinodermata—The deuterostome animal phylum of coelomate, marine, benthic (exceptionally pelagic) metazoans lacking a differentiated head or brain and typically having pentamerous symmetry, an endoskeleton formed of plates and other distinct ossicles composed of crystalline calcite, and a water vascular system; includes the familiar starfish and brittle stars (Asterozoa), sea urchins, sand dollars, and sea cucumbers (Echinozoa), sea lilies and related forms (Crinozoa), and lesser known living and extinct subphyla.
Echinozoa—The subphylum of echinoderms with radial symmetry, meridional pattern of growth producing an essentially globoid body, but no projecting appendages.
Ediacaran—The last period of the Neoproterozoic Era, ranging from about 635,000,000 to 543,000,000 years ago, followed by the Cambrian Period.
Edrioasteroidea—The class of sedentary, stalkless, radiate echinozoans that had a domal, clavate, or globose polyplated theca; ambulacra (generally five) endothecal, formed by floor plates and cover plates.
Embryophyta—The group of photosynthetic plants distinguished from the other plants by producing embryos that it nurtures within its body; includes mosses and all other land plants.
Endochondral bone—Cellular bone that develops by replacing cartilage with bone; typically found in the long bones of the skeleton.
Endoskeleton—A skeleton located within the body of certain organisms such as vertebrates, echinoderms, and various mollusks, for the purpose of muscle attachment, protection of internal organs, or both.
Epibiont—An organism that lives on and generally attaches to another organism.
Epifauna—Benthic fauna living on the substrate or on other organisms.
Eubacteria—Procaryotic organisms having a simple RNA structure and specific cell wall chemistry.
Eucaryota—The domain of organisms that contains cell nuclei and organelles; includes all organisms except eubacteria and archaeobacteria.
Euthycarcinoid—The enigmatic, mandibulate arthropod having a segmented tail region and up to twenty-eight pairs of undifferentiated walking legs; Middle Cambrian to Triassic.
Evaporitic—Pertaining to deposits of minerals precipitated when evaporation in a body of water exceeds the rate of introduction of additional water.
Exoskeleton—A stiffened external protective and supportive covering on certain animals, especially invertebrates.
External mold—Sediment that surrounds the outside surface of a shell or skeleton, that shows an impresson of the outside of the shell or skeleton, often showing external shell structures such as growth lines and ornamentation.
Facial suture—In trilobites, the suture bounding the adaxial margin of the eye surface; has anterior and posterior branches, each of which may extend over the border and onto the doublure; separates free cheek from fixed cheek.
Family—The rank in taxonomic hierarchy between order and genus.
Fenestrate—In bryozoans, a growth habit modification in which slender branches separated by rectangular open spaces called fenestrules are arranged in a reticulate pattern to form broad fronds.
Fixed cheek (=Fixigena)—In trilobites, the portion of the cranidium abaxial to the axial and preglabellar furrows and imaginary sagittal line bisecting the frontal area.
Foraminiferida—The phylum of amoeboid, unicellular protists that are generally microscopic, marine and include a chambered shell with foramen (perforations).
Free cheek (=Librigena)—In trilobites, the portion of the cephalon abaxial to the facial suture.
Fucoid—A member of or resembling the algae *Fucus*.
Fungus—The kingdom of single or multi-celled eucaryotes that typically have a system of filliments and spores; includes such organisms as mushrooms, yeasts, and molds; considered more similar to animals than to plants.
Funnel—See hyponome.
Gastropoda—A diverse class of vagile aquatic or terrestrial mollusks, including snails and slugs, that typically have a spiral calcareous

shell, although some shells are conical, cap-shaped or absent.
Genus—The rank in taxonomic hierarchy between family and species.
Geode—A cavity in a rock or fossil that has become lined with one or more layers of a mineral, generally silica or calcite.
Glabella—The central portion of the cranidium of a trilobite.
Gnathostomata—The superclass of craniates that have jaws; the jaws are derived from modified gill arches; this group includes the classes Placodermi, Chondrichthyes, and the grade Teleostomi, which in turn includes the classes Acanthodii and Osteichthyes.
Graptolithina—The class of sessile or pelagic colonial marine hemichordates that secreted a sclerotized exoskeleton with characteristic growth bands and growth lines; the thecae housing individual zooids were usually arranged in a single or double row along the branches (stipes) of the colony (rhabdosome); rhabdosomes originated by a single bud from the initial zooid, housed in a conical sicula, producing simple, branched or rarely encrusting colonies.
Hemichordata—The deuterostome animal phylum generally having embryonic pharyngeal openings but lacking the endostyle and notochord distinctive of the chordates.
Hold-fast—The root-like or disc-shaped structure at the distal end of a crinozoan's column used to anchor the animal to the substrate.
Holocene—The current epoch; in the Quaternary Period, following the Pleistocene Epoch, ranging from about 11,700 years ago to the present.
Hydraulic cement—Cement that chemically reacts and hardens with water, and is capable of doing so under water.
Hydrospires—Thin-walled, folded structures within the calyx of blastoids believed to have aided in respiration.
Hydrozoa—The class of generally carnivorous, colonial cnidarians having a tubular body, a vellum that aids in swimming in medusae, tentacles around their mouths that lack stinging cells, and may have either a polypoid stage, medusa stage, or both. The skeletons of colonial forms are generally chitinous, but calcareous in some species.
Hyolithida—The extinct group of probable lophophorates whose calcareous fossils include bilaterally symmetrical, conoidal sclerites that are open at their wider ends, an operculum, and a pair of lateral spines.
Hyponome (=Funnel)—In cephalopods, the fleshy tube used to propel the animal by forcing water through it.
Hyponomic sinus—In nautiloid and ammonoid cephalopods, the indentation of the shell's aperture through which the hyponome extends outside of the living chamber.
Hypostome—In trilobites, a plate attached to the ventral side of the carapace beneath the glabella; believed to have been related to feeding
Ichnogenus—A group of related ichnospecies.
Ichnospecies—The lowest rank in the hierarchy of trace fossil taxa.
Infaunal—Living within the sediment beneath a body of water.
Infraphylum—The rank in taxonomic hierarchy between subphylum and class.
Internal mold—Sediment filling the inside surface of a shell, often showing impressions of internal shell structures such as muscle scars.
Jurassic—The period in the Mesozoic Era between the Triassic and Cretaceous Periods, ranging from about 201,000,000 to 145,000,000 years ago.
Lake Church Formation—The Middle Devonian formation in southeastern Wisconsin that immediately underlies the Thiensville Formation and, depending on the exact location, overlies either the Waubakee Dolomite or Racine Dolomite.
Lancet plate—Elongate spear-shaped plate inserted between the deltoids of blastoids. The brachioles are attached to these plates.
Lindwurm Member—The member of the Milwaukee Formation between the Berthelet and North Point Members; more thinly bedded and fossiliferous than the Berthelet Member.
Linguliformea—The subphylum of brachiopods having organophosphatic stratiform shells that are not articulated or with rudimentary articulation not involving teeth and sockets, and possessing a pedicle formed as an extension of the ventral body wall.
Lophophorate—A group of lophophore-bearing coelomate organisms that includes brachiopods, bryozoans, and phoronids.
Lophophore—A ciliated structure used by certain aquatic animals, such as brachiopods, bryozoans, phoronids, and hyoliths, to gather and transport food to the mouth.
Lycophytina—Tracheophytes having leaves arranged in a dense, regular spiral pattern that result in distinctive scar patterns in the deciduous members of the group; leaves are narrow and typically with one unbranched vein; some Middle Devonian through Carboniferous forms were arborescent; lycopods.
Keratinous—Composed of the fibrous protein, keratin; the main constituent of hair, feathers, nails, hooves, and horns.
Kingdom—The rank in taxonomic hierarchy between domain and phylum.
Lycopod (=lycophyte)—A member of the Lycophytina
Malacostraca—The class of crustaceans with a carapace of widely varied form typically covering all or most of the head and trunk regions; head typically bearing paired compound eyes commonly located at tips of movable eye stalks, with five pairs of appendages behind the eyes (antennules, antennae, mandibles, maxillules and maxillae); trunk composed of eight thoracic somites and six or seven abdominal somites, most equipped with paired appendages; includes crabs, lobsters, and crayfish.
Mandible—The chitinous chewing structure found in millipedes, crustaceans, insects and certain other living and extinct arthropods; also the lower jaw of vertebrates.
Marcasite—The morph of the mineral iron sulfide (FeS_2) in the orthorhombic crystal system and is tin white in color on freshly fractured surfaces, but quickly weathers to ferrous sulfide, which is crumbly and orange in color.
Medusoid—Having the mode of life, form, orientation of a medusa, thus being free-swimming, having an umbrella-shaped body, with a mouth and tentacles facing ventrally.
Membrane bone—Bone that forms in membranous tissues and not as a replacement of cartilage.
Meridional—Pertaining to a meridian, which is a line running from pole to pole of a rounded structure; describes the pattern of growth found in echinoids and holothurians (sea cucumbers).
Mesenteries—In anthozoans, the vertically-oriented sheet-like partitions that radiate inward from the body wall toward the central cavity of the polyp; the organs of digestion, adsorption and excretion, and the site of development of the gonads.
Mesozoic—The era immediately following the Paleozoic and preceding the Cenozoic Eras, beginning with the Triassic Period and continuing to the Cretaceous Period.
Metazoa—See Animalia.

Microconchida—The order of encrusting tentaculitoids having a small, spiral shell.
Millerite—The mineral nickel sulfide (NiS); metallic pale brass-yellow crystals frequently forming as radiating needles associated with calcite; sometimes secondarily altered on its surface to emerald green zaratite (hydrous nickel carbonate).
Millipede—A member of the myriapod class Diplopoda, which is typically characterized by possession of an elongated, cylindrical body and numerous pairs of legs distributed as two pairs per segment.
Milwaukee Cement Company—A large producer of hydraulic cement located at the northeast side of Milwaukee, Wisconsin, and operating between 1876 and 1909. The source rock was certain layers (primarily the Berthelet Member) of the Milwaukee Formation.
Milwaukee Formation—A Middle Devonian (Givetian) formation located in Milwaukee County, Wisconsin, and lying above the Middle Devonian Thiensville Formation and below the Upper Devonian Antrim Shale; contains a diverse biota of fossilized invertebrates, fish, and rare land plants; consists of three members, in ascending order: Berthelet, Lindwurm and North Point.
Mollusca—The protostome animal phylum consisting of aqueous and terrestrial, nonsegmented (except for the Polyplacophora) invertebrates having a fleshy mantle that secretes a calcareous shell of one or more pieces. The shell may be exteriorly-facing, enclosed by the mantle, reduced or lost; includes snails, clams, squids and related forms.
Monophyletic—A relationship among a group of organisms in which the group is descended from a common ancestor, and includes all of the descendant groups.
Monticule—In some bryozoans, a cluster of polymorphs which makes a prominence on the surface of the zoarium; a macula.
Natural reef—A large subaqueous moundlike structure composed of the skeletons of various organisms such as corals, bryozoans and/or other biota that secrete calcareous skeletal materials; typically includes the living producers of these materials and a diverse and prolific community of other species that also inhabit the reef. Although the Milwaukee formation contained bryozoans and corals, it lacked reef structure.
Nautiloidea—The subclass of cephalopods having small to large external shells ranging in shape from straight to tightly coiled, simple, curved septa and sutures, and lacking the close-packed endocones or complex annular deposits characteristic of some of the other subclasses.
Nekton—Animals that swim freely against currents.
Neogene—The period in the Cenozoic Era between the Paleogene and Quaternary Periods, ranging from about 23,000,000 to 2,600,000 years ago.
Neoproterozoic—The era preceding the Paleozoic Era.
Nitrogenous (Biological context)—Containing nitrogen, such as chitin, skin and many other biologically produced materials.
North Point Member—The youngest member of the Milwaukee Formation, lying above the Lindwurm Member and below the Upper Devonian Antrim Shale; lithologically similar to and apparently continuous with the Lindwurm Member, but contains many silicified fossils.
Notochord—One of the defining features of the Chordata, consisting of a rod-like flexible embryonic structure that persists in the adults of certain subphyla and contributes to the ossification of vertebrae in the vertebrates.
Operculum—The keratinaceous or calcareous plate attached to the foot of many gastropods for partial or complete closure of the aperture.
Order—The rank in taxonomic hierarchy between class and family.
Ordovician—The period in the Paleozoic Era between the Cambrian and Silurian Periods, ranging from about 485,000,000 to 444,000,000 years ago.
Ossicles—The discrete calcareous entities that comprise the exoskeleton of echinoderms or are embedded in their skin, providing support, protection and defense.
Osteichthyes—The class of gnathostomes having endochondral or membrane bone; skull with sutures; lung(s) usually present, functioning either as air-breathing organs or as buoyancy-controlling swimbladders or both; includes the Sarcopterygii and the Actinopterygii
Ostracoda—The class of crustaceans that is aquatic, semi-microscopic, kidney bean-shaped, bi-valved, and mostly free-swimming.
Otoliths—Calcareous bodies within the inner ear of teleosts; play a critical role in the sense of balance.
Paleogene—The first period in the Cenozoic Era, between the Cretaceous and Neogene Periods, ranging from about 66,000,000 to 23,000,000 years ago.
Paleozoic—The era between the Neoproterozoic and Mesozoic Eras, beginning with the Cambrian Period and ending with the Permian Period.
Paraphyletic—A relationship among a group of organisms in which the group is descended from a common ancestor, but does not include all of the descendant groups; e.g., Pisces, which was defined to exclude the Tetrapoda, which is believed to have descended from the same common ancestor.
Pedicle—The fleshy organ of a brachiopod with which the animal typically attaches to the substrate.
Pedicle valve (=ventral valve)—Valve through which pedicle commonly emerges, usually larger than dorsal valve and invariably containing teeth when present.
Pedogenic—Pertaining to the formation and development of soils and related sediments.
Pelagic—The marine zone that includes the entire column of water above the benthic zone.
Perichondral bone—Bone that is typically thinly laminated, acellular, surrounds the soft tissues and is ossified from a layer of tissue that surrounds the cartilage as it develops.
Periderm—The thin, flexible organo-apatitic covering of a conulariid.
Permian—The last period in the Paleozoic Era, between the Carboniferous and Triassic Periods, ranging from about 300,000,000 to 250,000,000 years ago.
Permineralization—The fossilization process in which calcite or to a lesser extent silica or other dissolved minerals in the groundwater crystallize within the pores of skeletal remains
Phanerozoic—The eon following the Precambrian, ranging from about 541,000,000 years ago to the present.
Phoronida—The phylum of solitary lophophorate metazoans having a long, flexible tube-like body.
Phyllocarida—The order of malacostracan crustaceans whose head and thoracic regions are encased in a bivalved chitinous carapace.
Phylum (pl. a)—The taxonomic rank typically lying between the rank of kingdom and class or subphylum.
Pinnules—Fine projections of certain crinoid arms that may be used as filters during feeding.

Pisces—The name of a subdivision of vertebrates previously used to include what were considered to be fishes and to exclude tetrapods; generally abandoned due to the paraphyletic nature of the group.
Placodermi—The class of marine and freshwater gnathostome (jawed) craniates having a mostly cartilaginous endoskeleton and boney dermal plates covering their head and trunk.
Plankton—Animals and plants that float or drift in a body of water, but are unable to swim against the currents.
Polychaeta—The diverse class of mostly marine annelids that typically bear unjointed appendages that develop as extensions of the animal's body segments.
Polypoid—Having the mode of life, form, orientation of a polyp, thus typically being attached at its base, with a mouth and tentacles facing dorsally.
Pleistocene—The first epoch of the Quaternary Period, between the Pliocene and Holocene Epochs, ranging from 2,580,000 to 11,700 years ago.
Polymorphic (Biological context)—Having more than one form within the same species, each generally having its own function; e.g., honey bees with their drone, worker, and queen morphs.
Precambrian—The eon prior to the Phanerozoic, beginning with the formation of Earth (about 4,600,000,000 years ago) and ending about 541,000,000 years ago.
Prokaryota—The domain of single-celled organisms that lack a nucleus and organelles.
Protista—The kingdom of eukaryotic organisms that are typically microscopic and single-celled or in the form of colonies of identical cells.
Protostomia—The subkingdom of animals having a mouth that delelops from the first opening into the embryonic gut (blastopore); includes the Mollusca, Annelida and Arthropoda.
Ptyctodontiformes—The placoderm order characterized by its sturdy, blunt, frequently textured chewing surfaces presumably specialized for crushing shells.
Pygidium (pl. a)—In trilobites, the posterior portion of the exoskeleton separated from the thorax by articulation.
Pyrite—The morph of the mineral iron sulfide (FeS_2) that is in the cubic crystal system and metallic brassy yellow in color; weathers to limonite; "fools' gold."
Quaternary—The current geological period, following the Neogene Period, ranging from 2,580,000 years ago to the present.
Radiolaria—The phylum of protists bearing a siliceous test of varying shapes and intricate designs.
Radula—The keratinaceous, tongue-like structure in the mouth of most mollusks used for rasping food off a surface or drilling into shells.
Ramose—Branching, as in ramose corals or bryozoans.
Replacement (Paleontological context)—A type of fossil preservation in which groundwater-derived minerals entirely replace the skeletal materials of the organism.
Reptant—A growth habit of such colonial animals as certain corals and bryozoans, in which the colony is typically chainlike, branching and attached to the substrate by most or all of its habitation cells.
Rhabdosome—The sclerotized exoskeleton of an entire graptolite colony.
Rheophilic—Preferring or living in flowing water.
Rhombohedral cleavage—The tendency of minerals in the hexagonal crystal system to break in pieces having the shape of a rhombohedron.
Rugosa—The extinct order of anthozoan cnidarian corals that were sessile, marine, calcareous, generally cone-shaped, solitary, and distinguished from each other largely by features relating to their septa.
Sabkha—A coastal region subject to extensive flooding and evaporation, typically resulting in the deposition of salts; a salt flat.
Sarcopterygia—The subclass of Osteichthyes consisting of lobe-finned fish and tetrapods.
Schizochroal—Pertaining to a type of trilobite eye in which its visual surface consists of a number of biconvex lenses, rounded or polygonal in outline, each lens covered by an individual cornea and separated from others by sclerotic walls.
Schott—In conulariids, a transverse wall in the apical region of the periderm; may have been a repair structure produced in response to mechanical severance of the body.
Sclerite—A sclerotized structure or section of one that adds rigidity and may serve other functions in various animals, especially arthropods; examples include a cephalopod's beak and a lobster's exoskeleton, which is composed of many sclerites.
Sclerotized—Hardening of the cuticle of an organism, such as an arthropod's outer covering or the human fingernail, whereby a more rigid organic substance, especially sclerotin, is produced.
Scolecodont—The chitinous jaw of a polychaete worm.
Scyphozoa—The class of the Cnidaria whose members are marine and, unlike medusoid Hydrozoa, have stinging cells on their tentacles, lack a vellum, and typically have a four-partitioned body and medusoid adults.
Sea pen—A member of the anthozoan order Pennatulacea, named for their resemblance to quill pens.
Segment—The exoskeleton of a somite, generally of the thoracic region of an arthropod.
Septum (pl. a)—A calcareous partition between the chambers of certain cephalopod shells, such as nautiloids, the extinct ammonoids, and other cephalopods that bear an exterior shell. Also, in certain anthozoans such as rugose corals, a radially disposed longitudinal partition of a corallite.
Sessile—Attached to the substrate, incapable of self-propulsion.
Sicula—The skeleton of the initial zooid of a colony of graptolites.
Silica—The colorless mineral SiO_2, belonging to the hexagonal crystal system and one of the most abundant minerals in Earth's crust; includes the familiar quartz, chert, and other lesser known morphs.
Siliceous—Composed of or containing silica.
Silicified—A type of fossil preservation in which the skeletal remains are replaced by silica.
Silurian—The period in the Paleozoic Era between the Ordovician and Devonian Periods, ranging from about 445,000,000 to 419,000,000 years ago.
Siphuncle—In externally-shelled cephalopods, the long slender or thick tube extending through all chambers to the apex and consisting of soft and shelly parts.
Somite—A transverse division of the body of an arthropod.
Species—Generally the lowest rank in taxonomic hierarchy, below genus; may be subdivided into subspecies or varieties. Members

of the same species are normally capable of breeding and producing fertile offspring, in contrast to members of different species.
Sphalerite—The mineral zinc sulfide (ZnS), in the cubic crystal system, and often of a brown to black, though variable, color; frequently found in octahedral form and associated with other sulfide minerals and calcite.
Spionida—An order of tentacle-bearing polychaete worms.
Spiracles—Small openings on the summit of the calyx of blastoids through which it is believed water was passed for respiration.
Spire—In gastropods, the volutions of the shell past the first volution, generally described in terms of its relative height, with spire height generally increasing with successive volutions.
Sporangia—Structures within which spores are produced.
Stalk—An elongated structure between an eye and the rest of the body, as in malacostracans, or between the rest of the organism and the substrate to which it is attached, as in conulariids and crinozoans. See Column.
Stipe—In graptolites, one branch of a branched rhabdosome or the entire colony of an unbranched rhabdosome.
Stratigraphy—The study of the successions of layered rocks, what they reveal concerning changes that have taken place at that region, their use in correlating with successions in other regions, and helping locate mineral deposits of economic interest.
Stromatolites—Laminated dome-like, sheet-like or branching structures comprised of finely laminated calcareous layers formed by cyanobacteria, contributing to free oxygen production and reef-building since Precambrian time.
Stromatoporoidea—An enigmatic extinct group of colonial, calcareous major reef-builders in the Paleozoic Era consisting of laminae and pillars; may have been poriferans (sponges).
Suspension-feeder—An organism that feeds on materials that are suspended in the water.
Suture—In externally-shelled cephalopods, the line of junction of the free part of the septum and inner side of the shell wall.
Symbiosis—A relationship between individuals of two species of oganisms in which at least one species benefits; the relationship may benefit, harm or have no effect on the other species.
Tabula (pl. ae)—In rugose and tabulate corals, transverse partition of the corallite, extending to the outer wall or inner marginal wall.
Tabulata—The extinct order of almost exclusively Paleozoic, marine, calcareous, colonial, anthozoan cnidarians characterized by their exoskeleton of slender tubes crossed by many tabulae.
Taxon (pl. a)—Any unit used in the formal classification of organisms; e.g., kingdom, order, species, division, etc.
Tegmen—In crinoids, the adoral part of the theca above the origin of free arms or occupying the space between them; may be calcified or not.
Teleostomi—The grade of gnathostomes that includes the classes Acanthodii and Osteichthyes; have three pairs of otoliths.
Tentaculitida—The order of the Tentaculitoidea that had shells that were typically small (1–80mm), conical in shape, had transverse rings and, in some species, transverse and/or longitudinal striations—the transverse ones being growth lines.
Tentaculitoidea—An extinct group of enigmatic, generally marine animals that bore a conical or spiral calcareous shell and included the suborders Tentaculitida, Microconchida, and Cornulitida.
Terrestrial—Living exclusively on land.
Theca (pl. ae)—In graptolites, the sclerotized tube or cup enclosing any zooid of the rhabdosome other than the sicula. In crinozoans and echinozoans, the cup-shaped or globoid plated test of the animal enclosing the main part of the visceral mass; the cup (calyx) of a crinozoan, especially of a blastoid.
Thiensville Formation—The Middle Devonian formation that underlies the Milwaukee Formation.
Trace fossil—The fossilized markings or structures made by the activity of an organism on or within a substrate or another organism; the fossilized evidence of the locomotion, movements, industry, or other behavior of an organism, as opposed to the organism itself; includes, tracks, trackways, burrows, borings, resting traces, nests, feces, etc.
Tracheophytes—Vascular plants.
Triassic—The first period in the Mesozoic Era, ranging from about 250,000,000 to 200,000,000 years ago; followed by the Jurassic Period.
Trilobita—The common and extremely diverse subphylum of marine Paleozoic arthropods characterized by a calcite-impregnated exoskeleton divided longitudinally into three lobes.
Turbid—A condition of water in which particles are suspended, thus lowering clarity and light transmission.
Unconformity—A plane of contact between two successions of strata, representing a hiatus in the geologic record due to erosion and/or non-deposition of sediment.
Vagile—Free to move about.
Ventral valve (=pedicle valve)—Valve through which pedicle commonly emerges, usually larger than dorsal valve and invariably containing teeth when present.
Vermiform—Having the form of a worm.
Vug—A cavity in a rock typically formed by disintegration or dissolution of an embedded organism or other structure, and often partially or completely filled with a mineral of a different composition than the surrounding rock.
Water vascular system—In echinoderms, a hydraulic mechanism consisting of an assemblage of canals and associated specialized structures used to assist in feeding, locomotion, respiration and other important functions.
Whorl—One 360-degree rotation of a spiral shell such as that of many gastropods.
Zooecium—The skeleton of a bryozoan zooid, consisting of calcareous layers of zooidal walls and any connected intrazooidal calcareous structures.
Zooid—In graptolites, the soft-bodied individual inhabiting the theca. In bryozoans, one of the physically connected, asexually replicated morphologic units which, together with multizooidal parts and any extrazooidal parts present, compose a colony.
Zoarium (pl. a)—A colony of bryozoans.

About the Authors

Donald G. Mikulic

Dr. Donald G. Mikulic is a curator at the Weis Earth Science Museum in Menasha, Wisconsin. He retired as a Senior Paleontologist at the University of Illinois at Urbana-Champaign in 2017. Don received his Ph.D. in geology from Oregon State University in 1980. He received the Eugene Richardson Award from the Mid-America Paleontology Society in 2003. For 37 years he served as an expert on the Paleozoic geology and paleontology of Illinois along with the geology and history of its industrial minerals. As a specialist on the geology and paleontology of Silurian rocks, especially in the Milwaukee and Chicago area, Mikulic conducted research projects in North America and Europe, and he worked as a curator at the Thomas A. Greene Memorial Museum at University of Wisconsin-Milwaukee (UWM), a consultant for the Milwaukee Metropolitan Sewerage District and the Wisconsin Department of Justice and is a steering committee member of the Niagara Escarpment Resource Network. Don has published numerous journal articles, abstracts and books, and has been collecting Milwaukee Formation fossils since the late 1950s.

Joanne Kluessendorf

Dr. Joanne Kluessendorf was an educator, women's advocate, founding Director of the Weis Earth Science Museum (WESM), recipient of numerous awards and appointments to various professional and scientific organizations, author of over 150 scientific publications, and a major voice in the preservation of significant geological treasures. A native Milwaukeean, Joanne was instrumental in establishing five National Historic Landmarks in Wisconsin, and had previously served as curator of the Thomas A. Greene Memorial Museum as well as the Museum of Natural History at the University of Illinois at Urbana-Champaign. The wife of Donald Mikulic, Joanne sadly passed away after a long illness before this book was completed. She was an inspiration to everyone with whom she came in contact. It is with great pleasure that we dedicate this book to her.

Kenneth C. Gass

Kenneth C. (Chris) Gass has been studying paleontology and collecting fossils independently in Wisconsin for over fifty years. He earned his B.A. degree in Geological Sciences at UWM in 1979. His initial area of research was Silurian trilobites of the family Encrinuridae, about which he published several taxonomic studies from 1977 to 1999. His focus shifted to trace fossils from the Cambrian tidal flats of Blackberry Hill in central Wisconsin when he wrote several articles and a book on the subject and popularized the related discoveries in talks, videos, magazine articles, and a public radio appearance. One of the articles that he co-authored won Honorable Mention for Best Paper Award from the *Journal of Paleontology*. Chris also developed a keen interest in the Devonian Milwaukee Formation. His field work included making hundreds of trips to Estabrook Park in Milwaukee to study that formation and its fossils before finally co-writing a book about it.

Carlton E. Brett

Professor Carlton E. Brett earned his Ph.D. at the University of Michigan in 1978, and that year accepted a faculty position in Geology at the University of Rochester. He joined the University of Cincinnati Geology faculty in 1998 and in 2014 was awarded the title of University Distinguished Research Professor. Dr. Brett's research lies at the interface between paleontology and sedimentary geology. He is pursuing studies relating regional and global changes in climate, sea level, and the carbon cycle to episodes of biotic change (bioevents) and extinction, as well as unusual sedimentary deposits, primarily in the middle Paleozoic. Carl also maintains active research interests in taphonomy (fossil preservation), paleoecology, especially the evolution of organism interactions such as predation, paleobiology of invertebrate animals, especially crinoids and trilobites, and cyclic stratigraphy. Carl is author or co-author of more than 250 scientific papers, 80 guidebooks, and five books. He was honored in 2010 with the University of Cincinnati's Rieveschl Award for research and has received several other honors, including an Alexander von Humboldt Research Prize in 2006, the 2008 Digby McClaren Medal for Lifetime Achievement in Stratigraphic Paleontology, the 2012 Raymond C. Moore Medal (Outstanding Paleontologist) from the Society of Sedimentary Geology, and the 2013 American Association of Petroleum Geologists Outstanding Educator Award.

Index to the Species, Subspecies, Unnamed Species, and Ichnospecies Cited in the Text
(For page numbers in bold print, an image of the indicated taxon is shown on that page.)

Also by Kenneth C. Gass

Siri Scientific Press (2015) 978-0-9929979-5-3 RRP £17.50
96 pp, 210 x 148 mm, soft cover,
81 colour illustrations, mainly photographs
Available at: www.siriscientificpress.co.uk

Reviews

Newsletter of the Paleontological Society (Priscum, Fall 2017, pages 19–20): "The photographs of the traces are extraordinary in their definition and clarity, and bring to life most vividly the activities of the animals. These trace fossils are of the greatest importance, as they are direct evidence that the beginnings of animal life leaving the sea for life on land began during the 540-500 million years of the Cambrian period, and not many millions of years later during the Silurian and Devonian periods."

Newsletter of the Palaeontological Association (2016, number 91, pages 107–108): "To cite the author from his first line of the preface to the book: 'A picture book on Blackberry Hills was inevitable. The place is simply too intriguing to remain buried in scientific journals…' I couldn't agree more." ... "... an excellent introduction for readers who want to dig deeper into the topic and the technical literature. Based on this I would recommend this book to all with an interest in natural history, scholars as well as amateurs."

Fossil News (Fall 2016: 51, abridged snippets): "...a beautiful, extensively illustrated book on one of the more important Middle Cambrian trace fossil deposits in the United States (the book contains 81 high-quality figures, most of them in colour and published only in this volume). ...a great addition to the library of anyone interested in trace fossils, Cambrian life, or the early invasion of land. For professionals, the book offers excellent images of the trace and body fossils of Blackberry Hill, as well as an extensive reference list. For amateur paleontologists, it provides a nice introduction to the importance of trace fossils in the fossil record (and includes a thorough terminology section, which helps with understanding the geological, ichnological and biological terms used throughout the text).

Niles Eldredge, evolutionary biologist: "A mystery revealed, this book charmingly tells the story of how clever scientific detective work has answered the question of what those first animals to visit the subaerial environments of our world really were – however fleeting their visits from their ancient ocean habitats may have been. Engagingly told by Chris Gass, one of the key participants, the story reveals the importance of a highly special area that holds the fossilized evidence that shows what really happened."

Gregory D. Edgecombe, The Natural History Museum (London): "The early history of life on land is a conundrum that has perplexed some of the foremost figures in the history of paleontology since the era of Richard Owen and Charles Darwin over 150 years ago. At centre stage in this mystery are trackways on sedimentary rocks that we now know are of Cambrian age - some 500 million years old - and were made by animals making the first excursions out of water in life's history. Here, Chris Gass weaves together the history of science and groundbreaking research conducted over the past 20 years to reveal the identities of these ancient trace makers. His book is a scientific detective story in words and photographs."

From the back cover

Solving the Mystery of the First Animals on Land: The Fossils of Blackberry Hill is the first book to cover the unique group of rock outcrops in central Wisconsin that appears to have put an end to a 150 year-old mystery of global interest. Since the mid-1800s, fossilized trackways and footprints have been found on beach deposits that date back to the Cambrian Period, some 500 million years ago in what is now North America – but fossils of the animals that made them were reluctant to reveal themselves. Thanks to Blackberry Hill, the identity of some of the first animals to walk on land is a secret no more.

Numerous color photographs of spectacularly preserved tidal flat trackways, animals, and parts of the habitat itself, all set in stone, help to tell the story of some of the first animals to explore this strange, new, and presumably hostile environment. Current interpretations made possible by the fossil discoveries are presented, including how some of the trackways were made and what might have lured the animals ashore. Peculiar, winding trackways from giant, slug-like organisms and other boneless animals are shown covering entire surfaces. This book also reveals other surprises discovered at Blackberry Hill, including body and trace fossils of another tidal flat dweller never before found in rocks as old as these, and the first evidence showing that large jellyfish were abundant in Cambrian times and were already subject to mass strandings. In this book, Kenneth (Chris) Gass thus demonstrates why Blackberry Hill holds a special place in the history of life on Earth.